CREWE'S ROLLS-ROYCE FACTORY

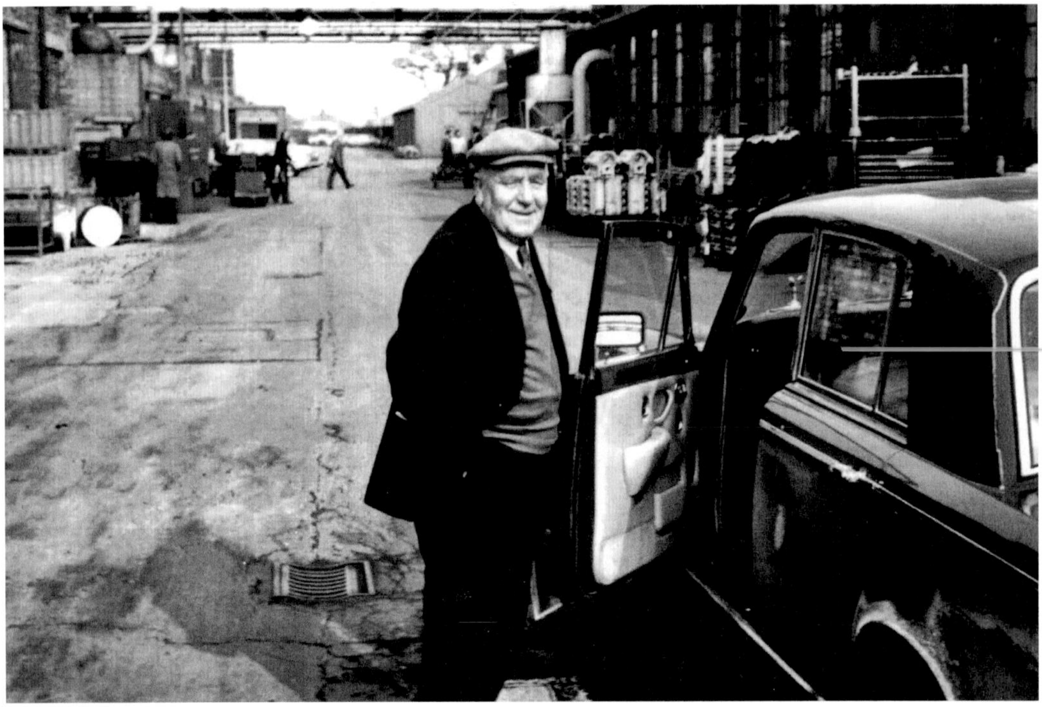

Arthur Payne, a labourer in the toolroom, alongside a Rolls-Royce Silver Shadow in the late 1970s. Looking south towards Pym's Lane.

CREWE'S ROLLS-ROYCE FACTORY

From Old Photographs

PETER OLLERHEAD & TONY FLOOD

AMBERLEY

This book is dedicated to all who worked at the Pym's Lane factory between 1938 and the end of the twentieth century.

First published 2010

Amberley Publishing Plc
Cirencester Road, Chalford,
Stroud, Gloucestershire, GL6 8PE

www.amberley-books.com

British Library Cataloguing in Publication Data.
A catalogue record for this book is available from the British Library.

ISBN 978 1 84868 859 9

Typesetting and Origination by FONTHILLDESIGN.
Printed in the UK.

CONTENTS

ABOUT THE AUTHORS

Peter Ollerhead worked at Pym's Lane after leaving school in 1952 as a centre lathe turner then a detail draughtsman. Sixteen years later a career change took him into the teaching profession where he remained for the next thirty years during which time he continued his interest in local history and second-hand books. He is the author of four other volumes on the town and people of Crewe.

Tony Flood worked for Rolls-Royce for forty years after joining the firm as an apprentice in 1963. For most of this time he worked in the Quality Engineering Department then latterly in preparing cars for demonstrations. His expertise has often been used by the Rolls-Royce Enthusiasts' Club especially as a judge at car rallies and similar functions.

PREFACE & ACKNOWLEDGEMENTS

Except for chapter eight I have deliberately used the headings from my previous book about Crewe Rolls-Royce which roughly follow various historical phases of the firm in Crewe.

Many of the persons thanked previously I thank again for their assistance. Firstly, I owe a great debt to the Rolls-Royce Enthusiasts' Club at Paulerspury for allowing me access to the negatives and slides in their archive. Without such help this book would not have been possible. Following on I must heartily thank Tony Flood, my co-author, for without his knowledge and cheerful assistance much of the material at Paulerspury would have remained unidentified. I also acknowledge help from Andrew Turner and Brian Kennedy for turning slides and film strips into images fit for publication and to Joan Elliott, Bernard Preston, Reg Spencer, Dorothy Pointon, Harry Jones, John Epps, Walter Lea, John Pennington, Sir David Plastow, Derek Coulson, Alan Hillier, Martin Bourne, Bob Tomkinson, Billy Consterdine and the late Alex Henshaw for the loan or gift of photographs. Susan Chambers provided expert guidance to Rolls-Royce material in both the Chester Record Office and the Crewe Heritage Centre so I offer her and the two organisations my appreciation. Finally I thank my wife who for almost fifty years has shown me great patience in addition to sharing my interest in things historical.

INTRODUCTION

When, in June 1938, it was made known that Rolls-Royce (RR) was setting up a factory in Crewe the pages of the *Crewe Chronicle* were filled with the news. Until that date Derby was the sole centre of RR production in England. In 1937 the Board of RR realising that the works at Derby could not cope with the flood of government orders for their aero engines in addition to manufacturing motor cars decided to construct a second factory away from Derby. This move was generated by the government's rearmament policy, an answer to the threat of Hitler's expansionist ambitions on the continent of Europe.

The pressing need for aero engines to power military aircraft was the only reason that RR came to Crewe. Without the rise of Hitler's Nazi Party the railways would quite probably have remained without a competitor for labour in south Cheshire. Hives and the Rolls-Royce Board chose Crewe and by the end of May 1938 a sixty acre site, in the west of the town, was purchased by the government enabling work to begin on a factory. Initially, seven men arrived from Derby to organise the manufacture of Merlin aero engines in as short a time as possible.

Between 1939 and 1945 around 26,000 Merlins covering 18 different marques were dispatched from Pym's Lane in addition to 2,000 Griffon power units. The number of employees rose rapidly from the Derby seven in July 1938 to almost 10,000 by 1944. A rough estimate would suggest that over 20% of these were women.

one

MERLINS MIGRATE TO PYM'S LANE

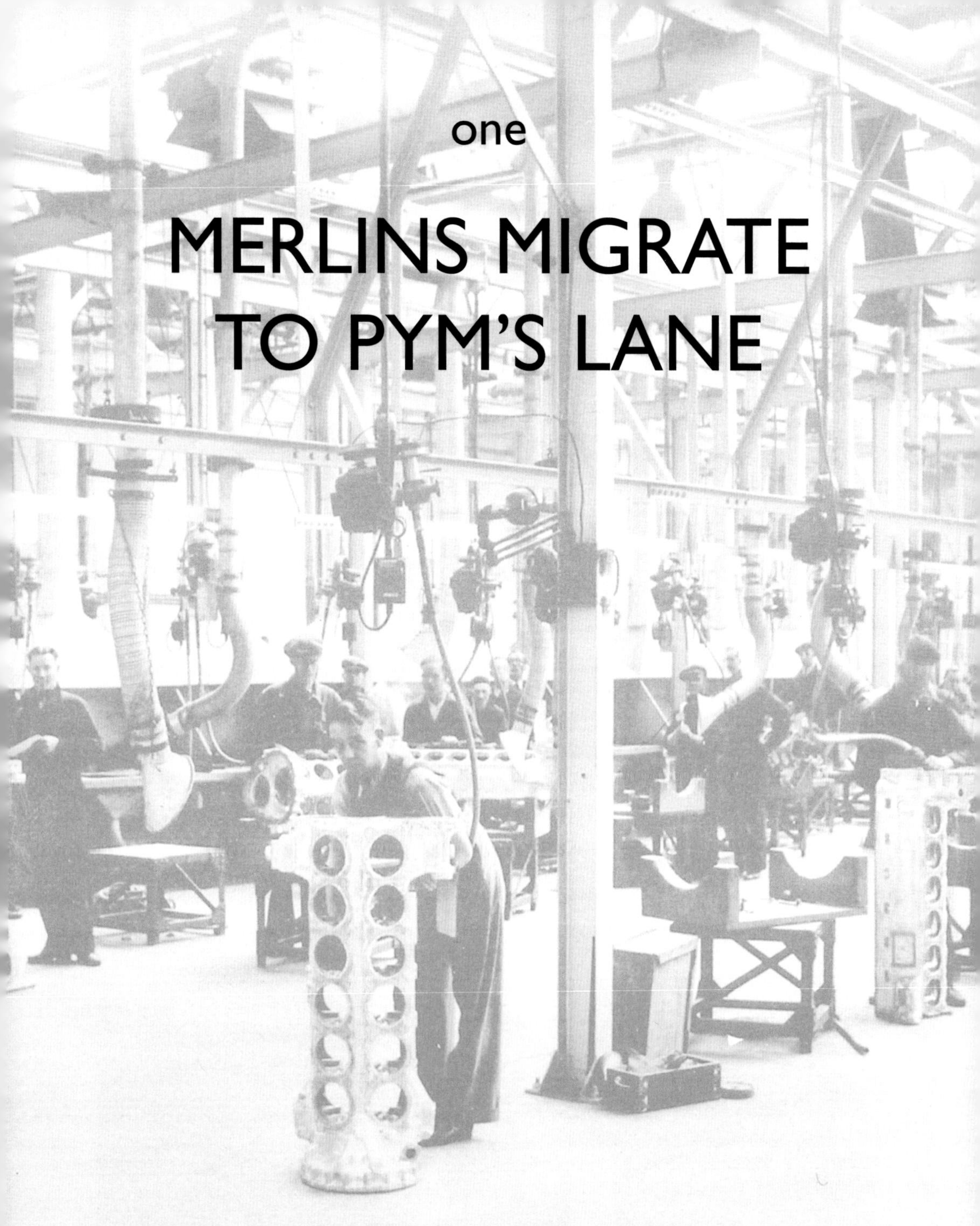

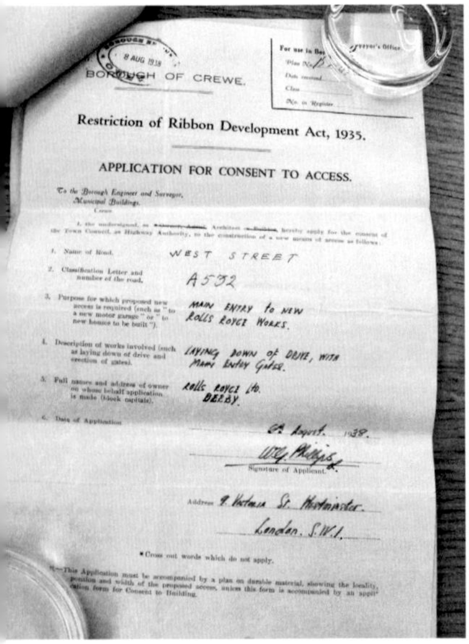

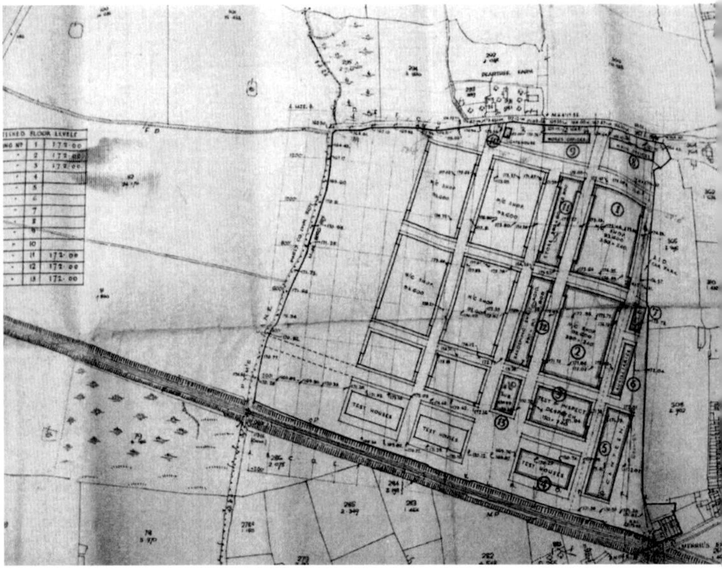

Above left: One of many notices of intent to build a factory for Rolls-Royce at Crewe submitted by the architect W. Phillips.

Above right: The first plans of the factory positioned the canteen near to Merrill's Bridge gate not fronting Pym's Lane which is where it was eventually sited. Notice also Pym's Lane turned southwards following the line of what is now Sunnybank Road.

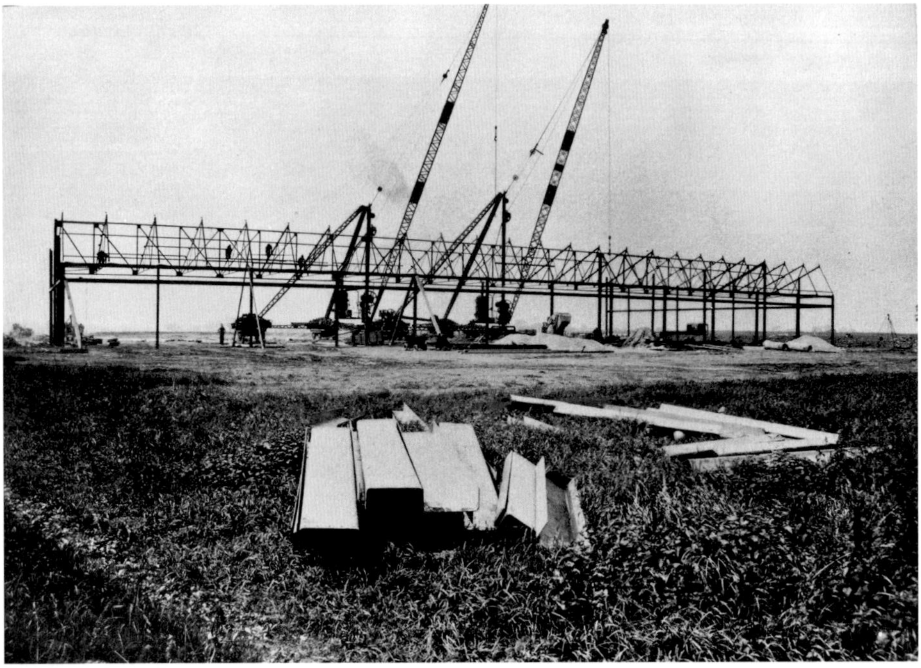

Steelwork at the north end of the main shop, 2 August 1938.

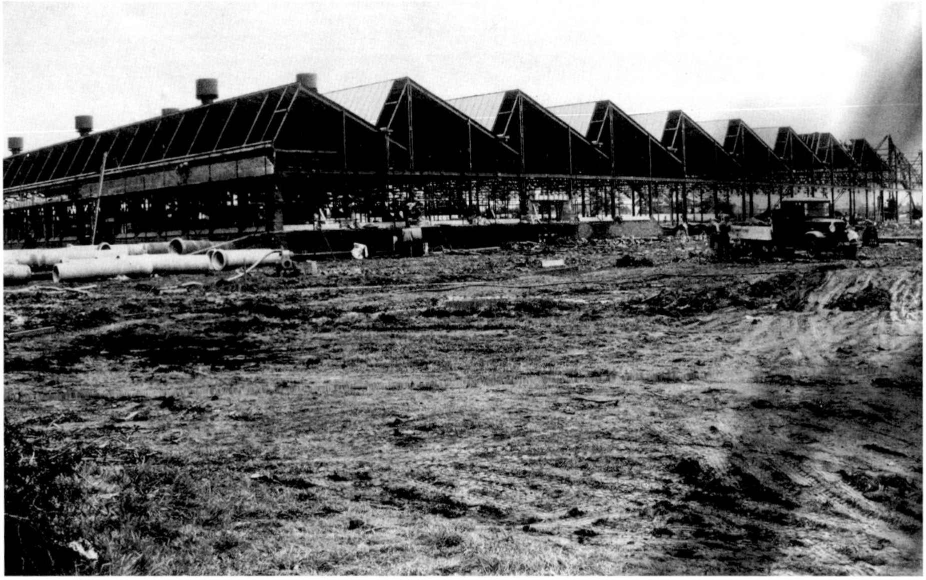

30 August looking east. Robertson roof ventilators that gave two changes of air every hour signalled the completion of the first bay.

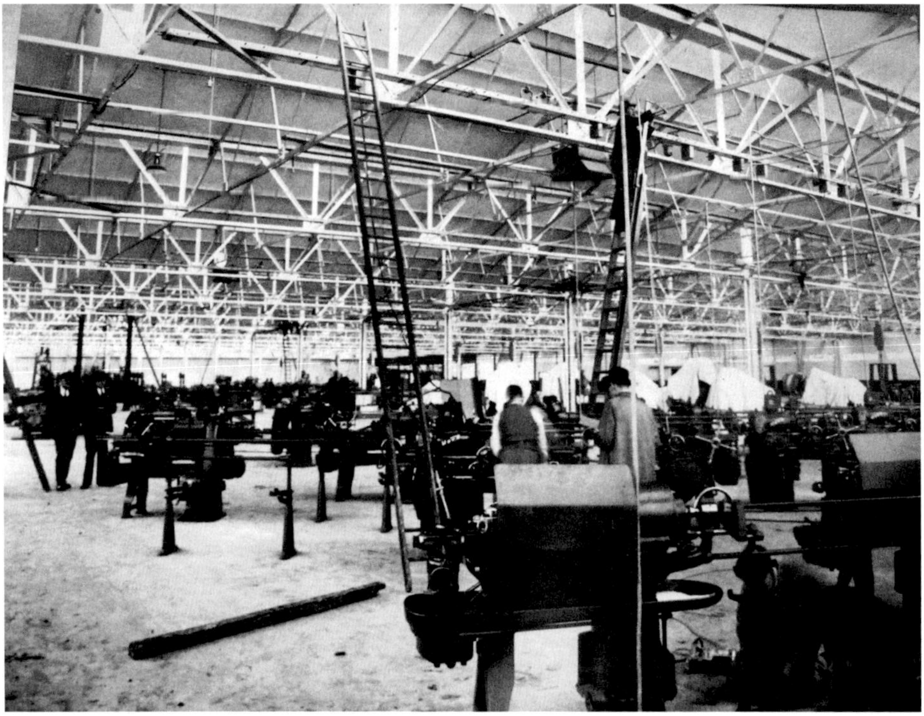

As soon as sufficient space was available capstan and bar auto lathes were purchased and located in the space that eventually became 101, 102 and 121 departments. A partition wall has been erected to allow machining to commence while building continues.

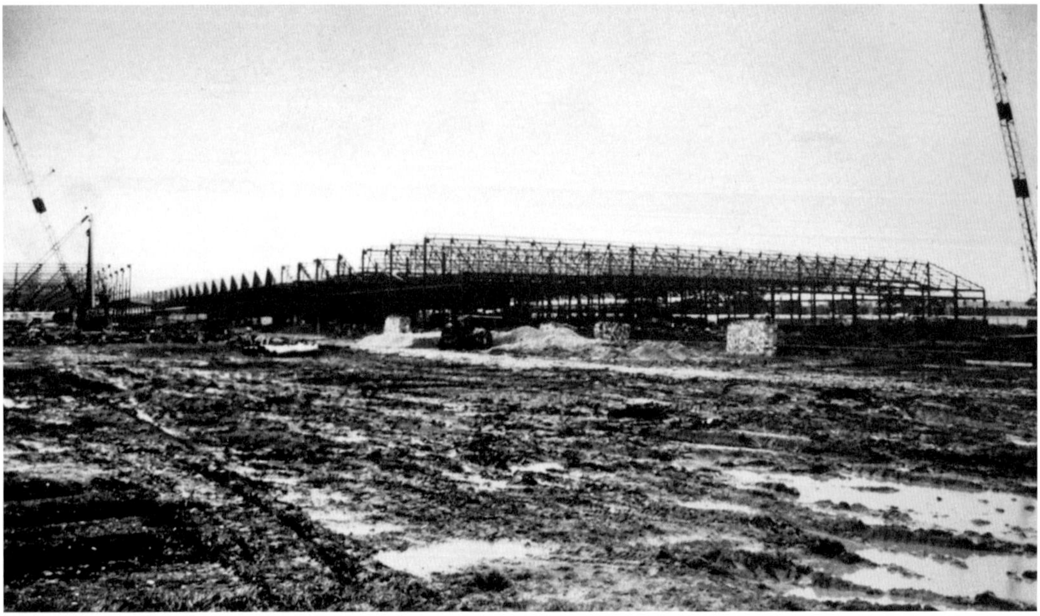

The main shop, looking north, on 18 October 1938, ten weeks into the project with around 600 persons on the payroll. This was also the date when machining of Merlin parts commenced in the north end of the building.

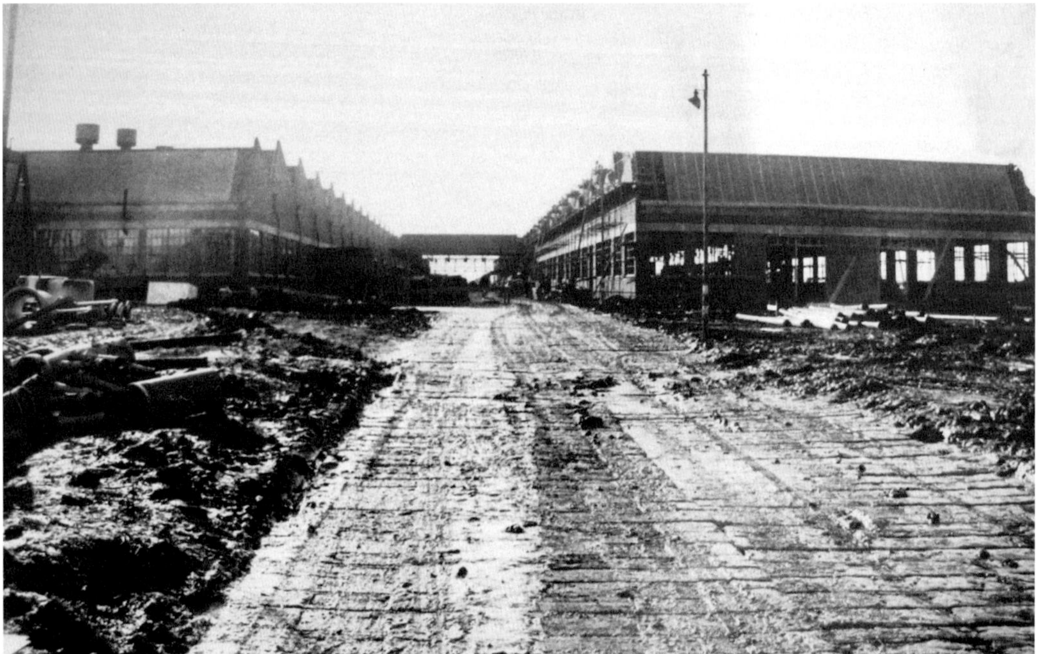

Looking south on 1 November 1938. The road in the centre, later called Second Avenue, was made from railway sleepers until the first phase of building was completed.

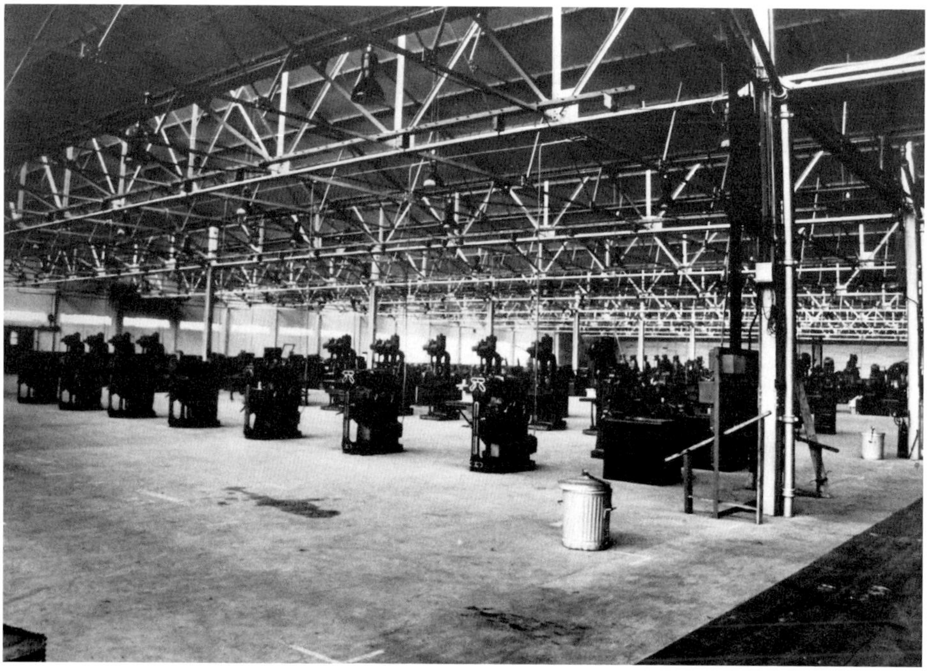

Vertical drilling machines in the main shop on 1 December 1938.

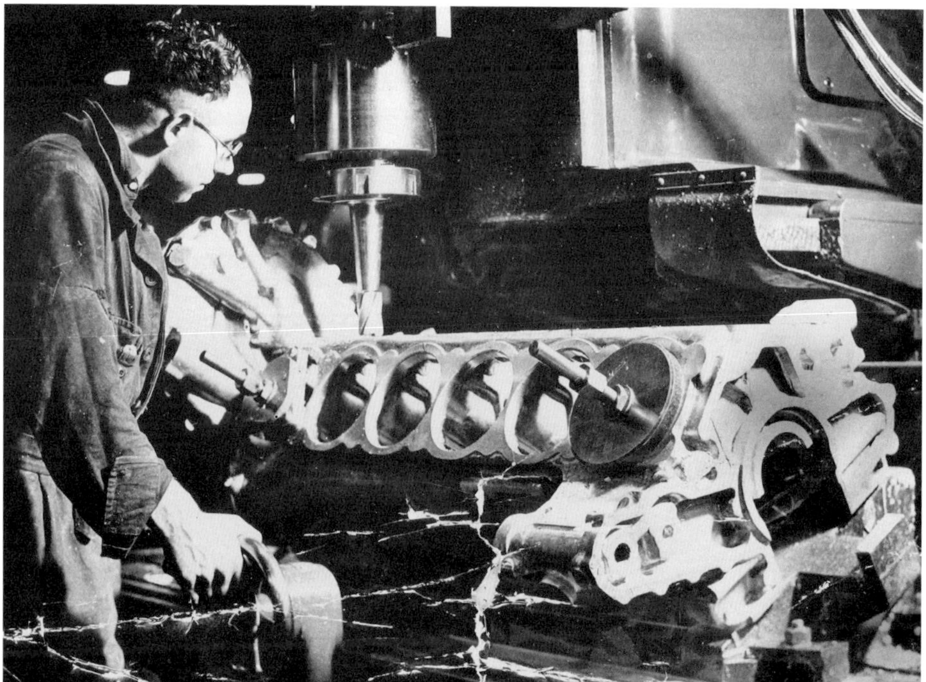

Milling a Merlin crankcase early in December 1938. The first superintendent of machining was J. Shaw assisted by B. Bliss.

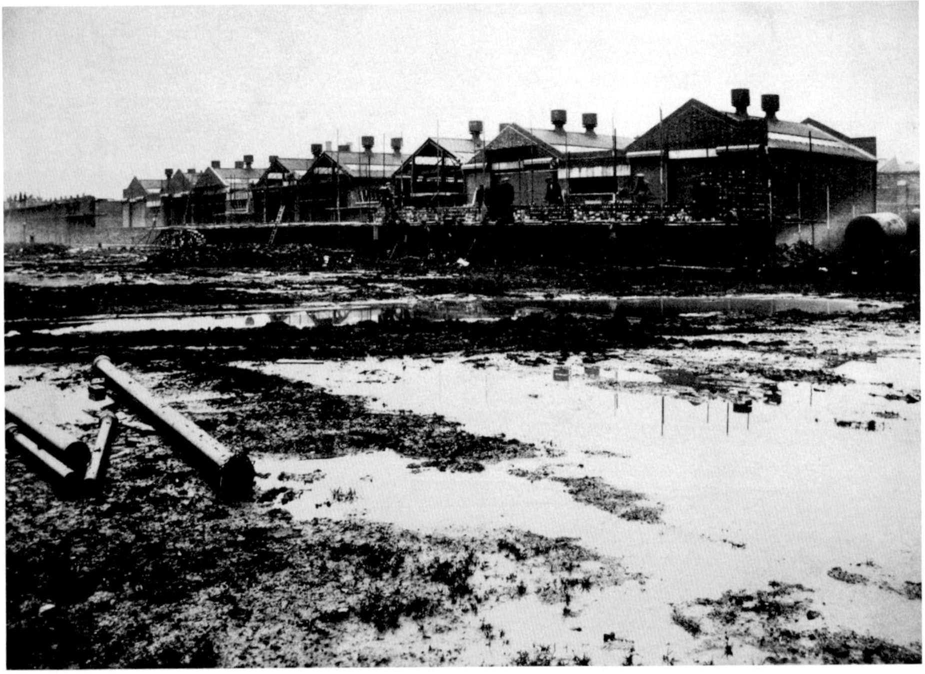

Late January 1939 looking south. The effects of the constant rain and snow on the heavy clay can still be seen, though the test houses at the southern extremity of the site are almost finished.

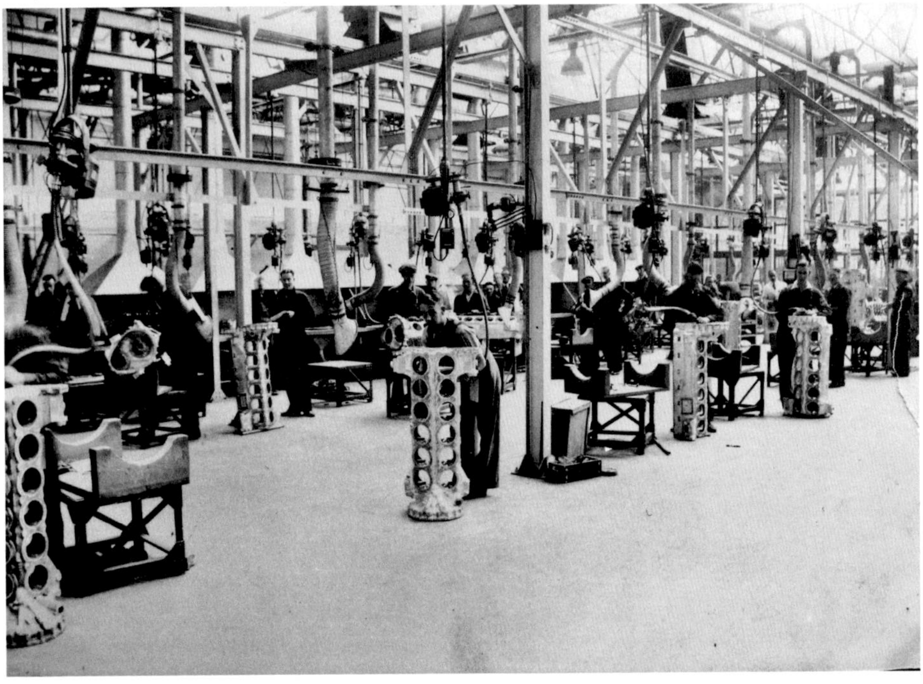

Sub-assembly prior to main Merlin build.

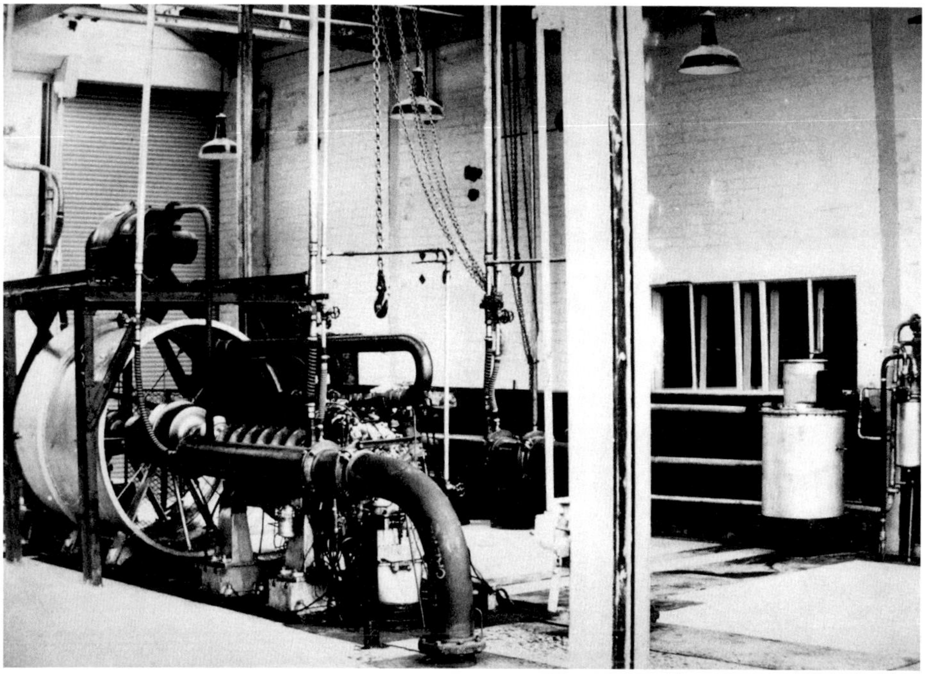

20 May 1939 was the date when the first Crewe manufactured Merlin engine was tested. Dick Garner, formerly riding mechanic for Kaye Don in the *Miss England* project, was chief tester.

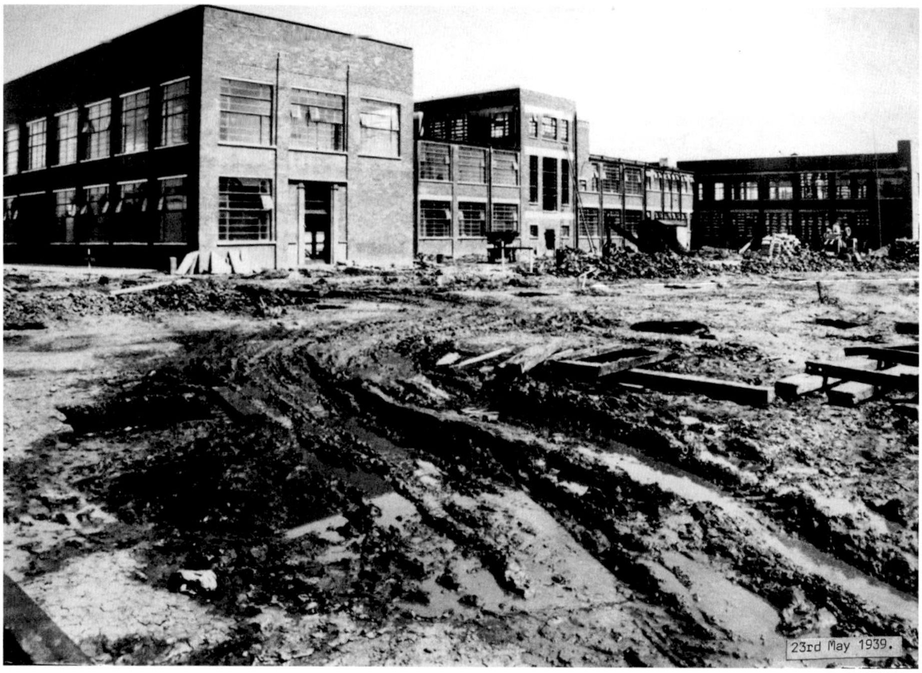

Looking north at the rear of the main office block on 23 May 1938. The ground floor windows nearest to the camera belonged to the personnel department.

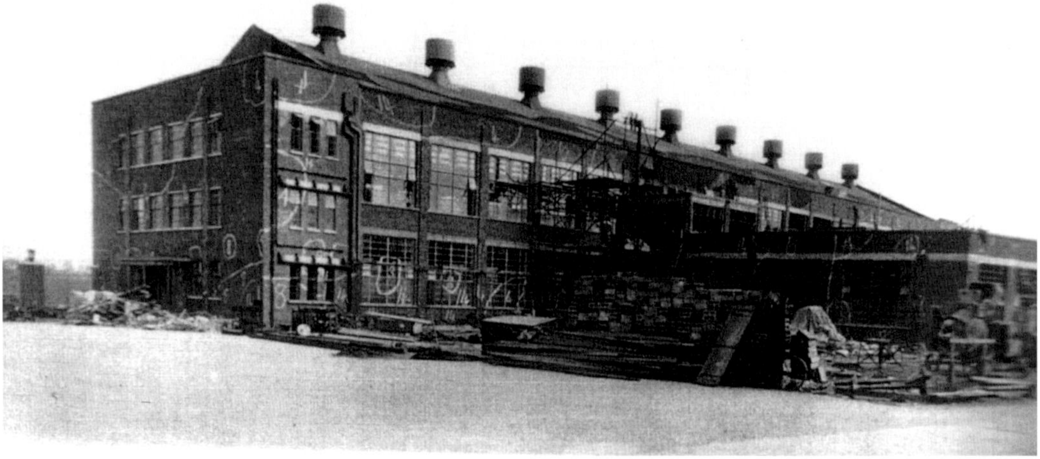

The canteen filled the lower storey of this block which was marked out for camouflage paint in late August 1939. An additional £14,000 was allocated for ARP measures.

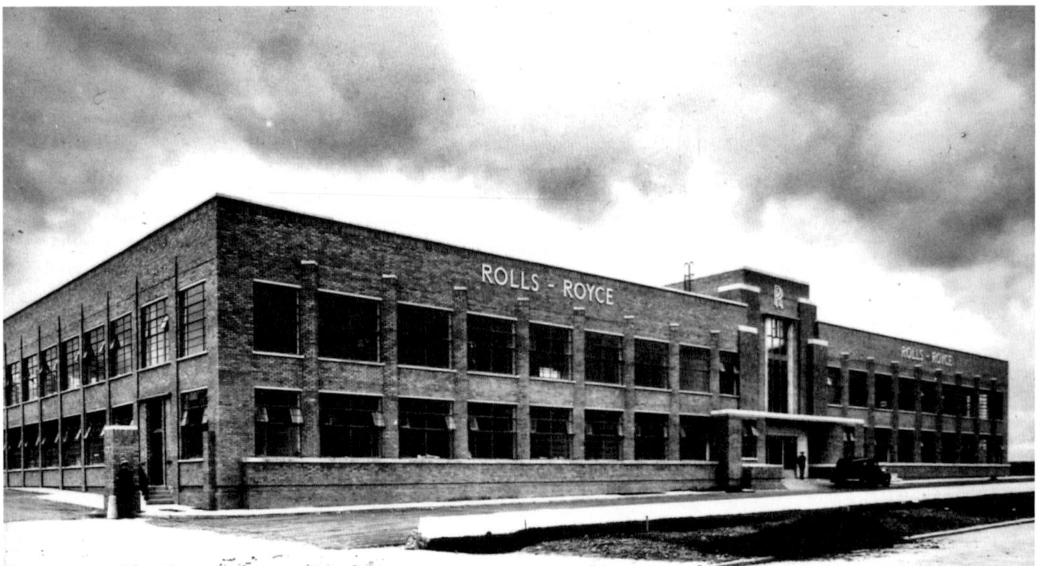

The main offices fronting Pym's Lane where for a short period an anti-aircraft gun was positioned on the roof.

two

THE MERLIN YEARS

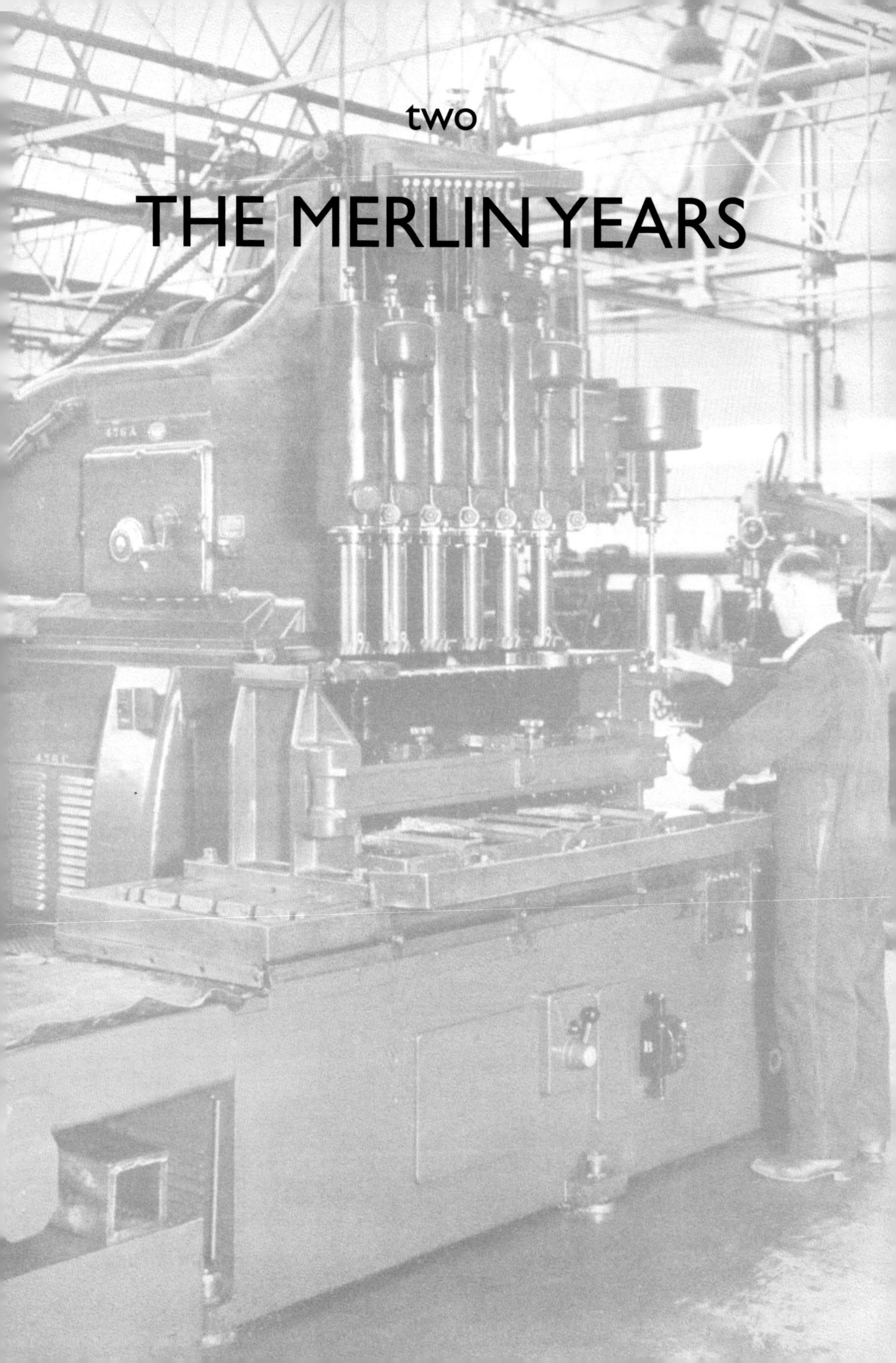

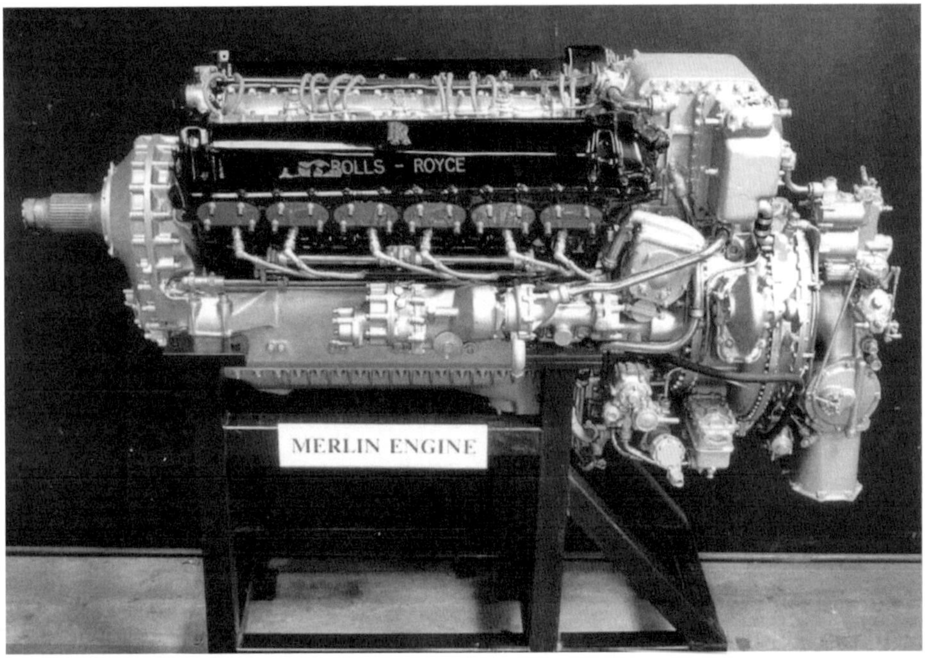

A Rolls-Royce Merlin was a twelve cylinder engine which ran at 3,000 rpm. Development increased its power from around 1000hp in 1940 to 2000hp four years later.

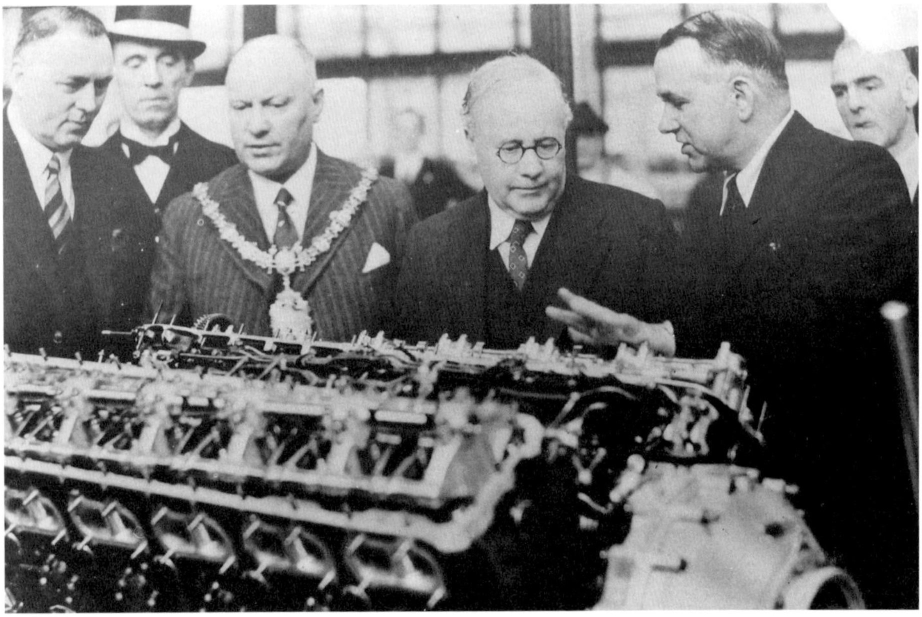

Sir Kingsley Wood, Air Minister in Chamberlain's government officially opened the Pym's Lane factory in July 1939. Ernest Hives, the driving force of all RR wartime endeavours is on the extreme right with John Morris, general manager at Crewe on the left. Harry Bricker was the mayor that year.

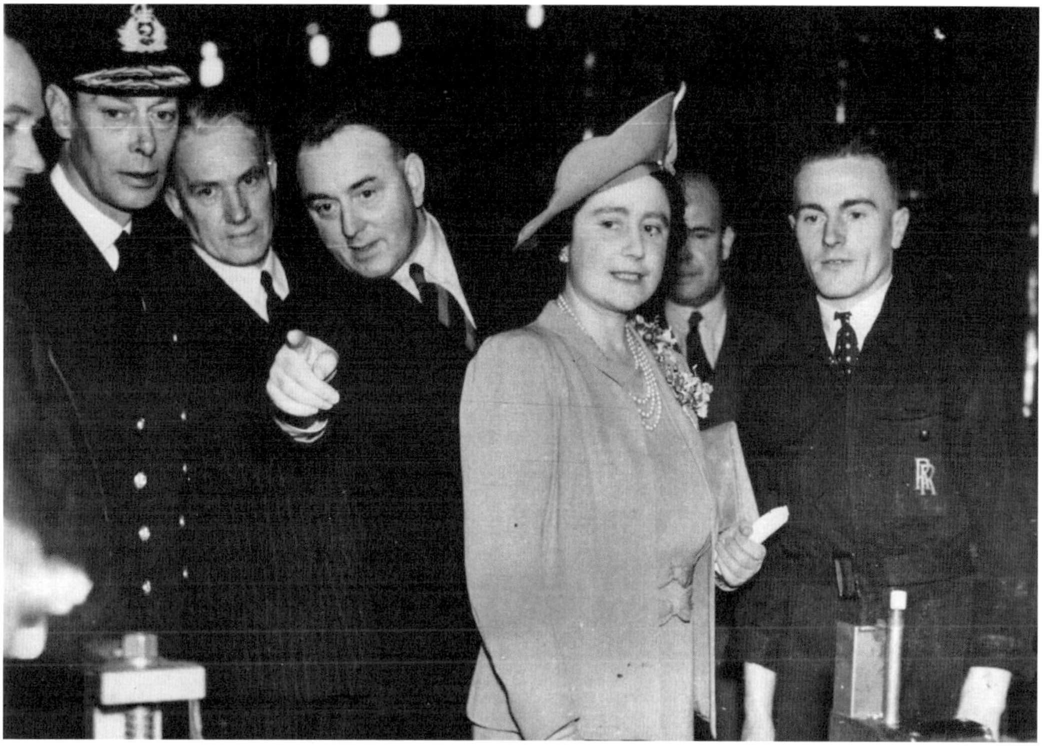

Above: King George VI and Queen Elizabeth made a visit to Pym's Lane in August 1940. John Morris is pointing to something of interest.

Right: A six spindle miller machining parts for the Merlin engine (*c.* 1939).

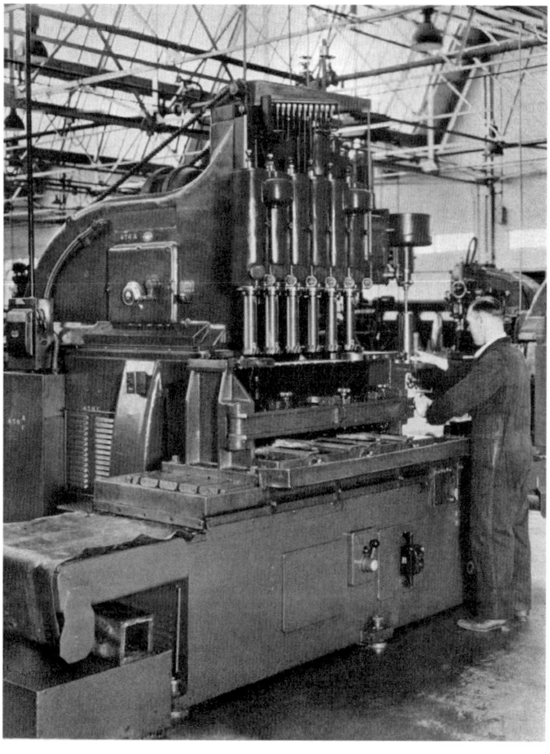

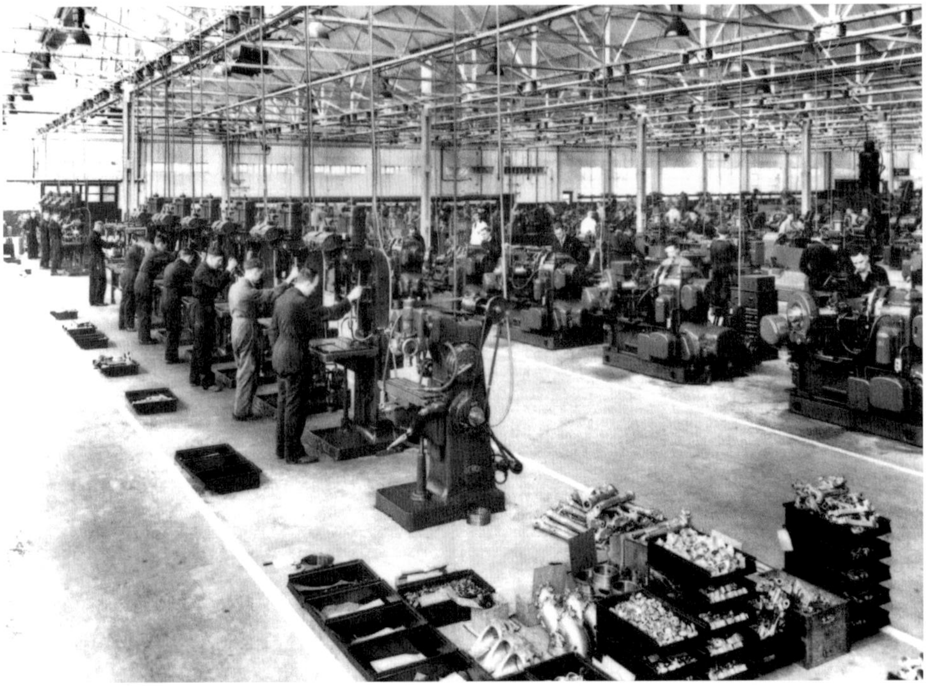

Vertical drilling machines in the north end of the main shop.

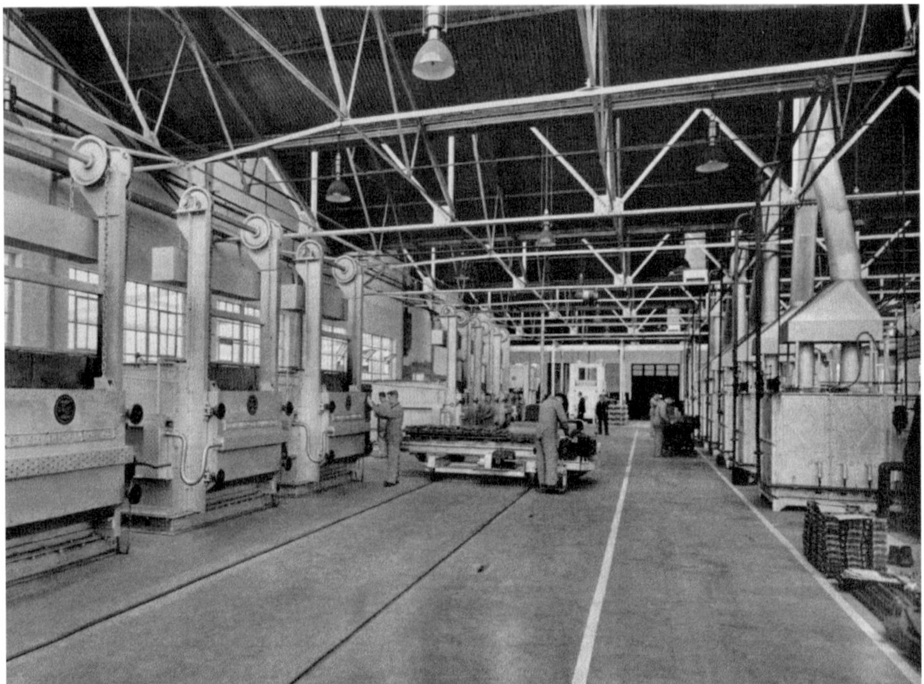

Heat treatment department, June 1939. Electric carburising furnaces and charging machine to the left with gas fired furnaces on the opposite side.

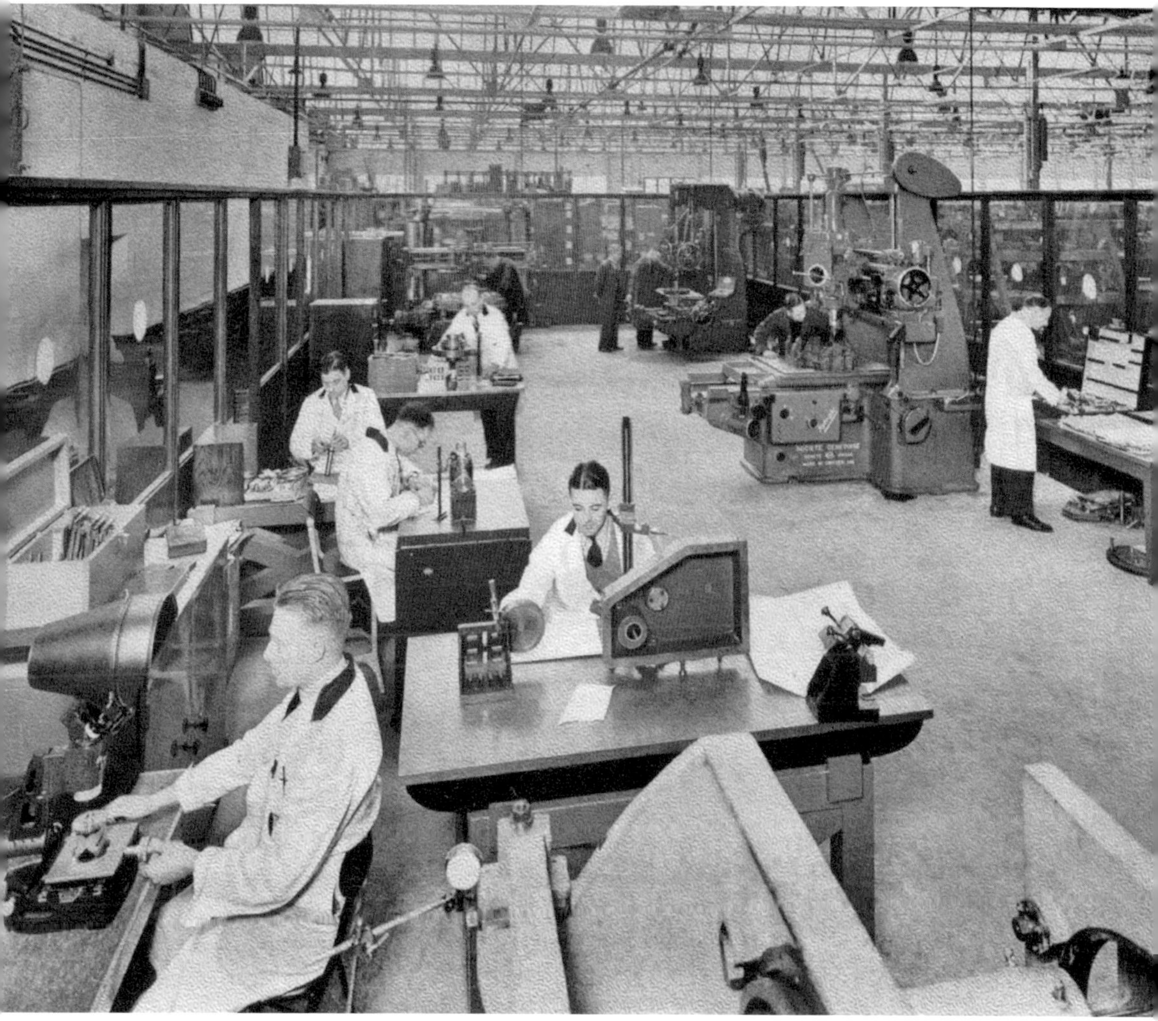

Jig boring and jig and tool inspection. Hives' policy was for Crewe to be more reliant on jigs and special tools compared with Derby.

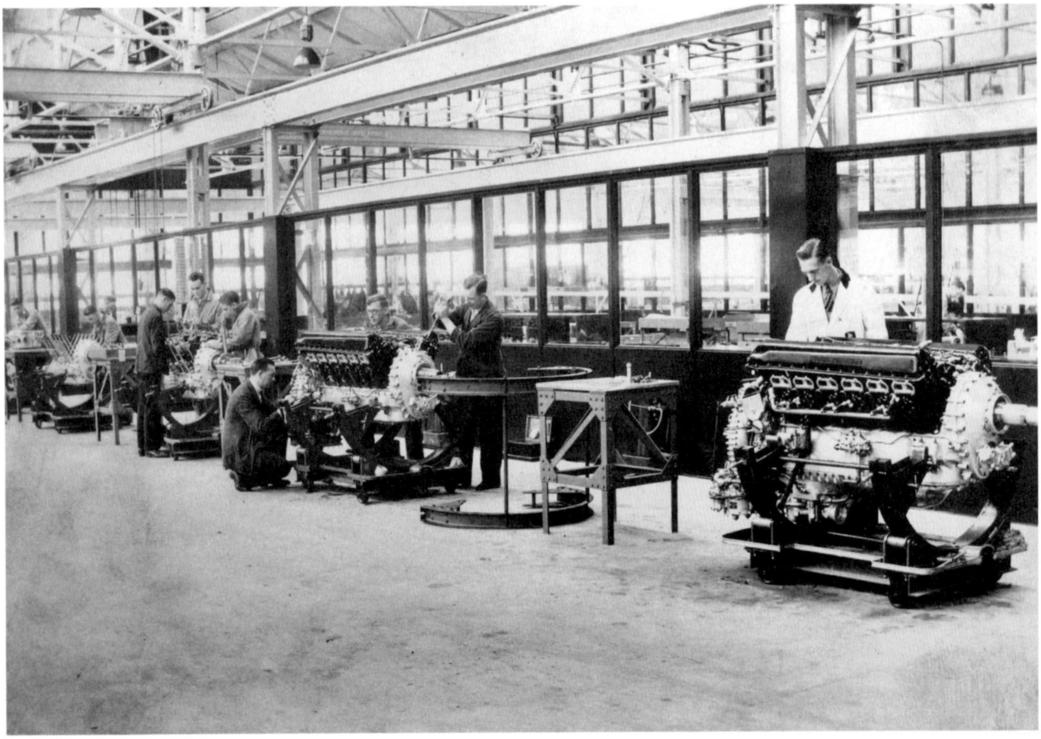

Merlin engine assembly late 1939. The first Merlin manufactured at Crewe was dispatched on 10 June 1939. By December the monthly rate reached 190 units.

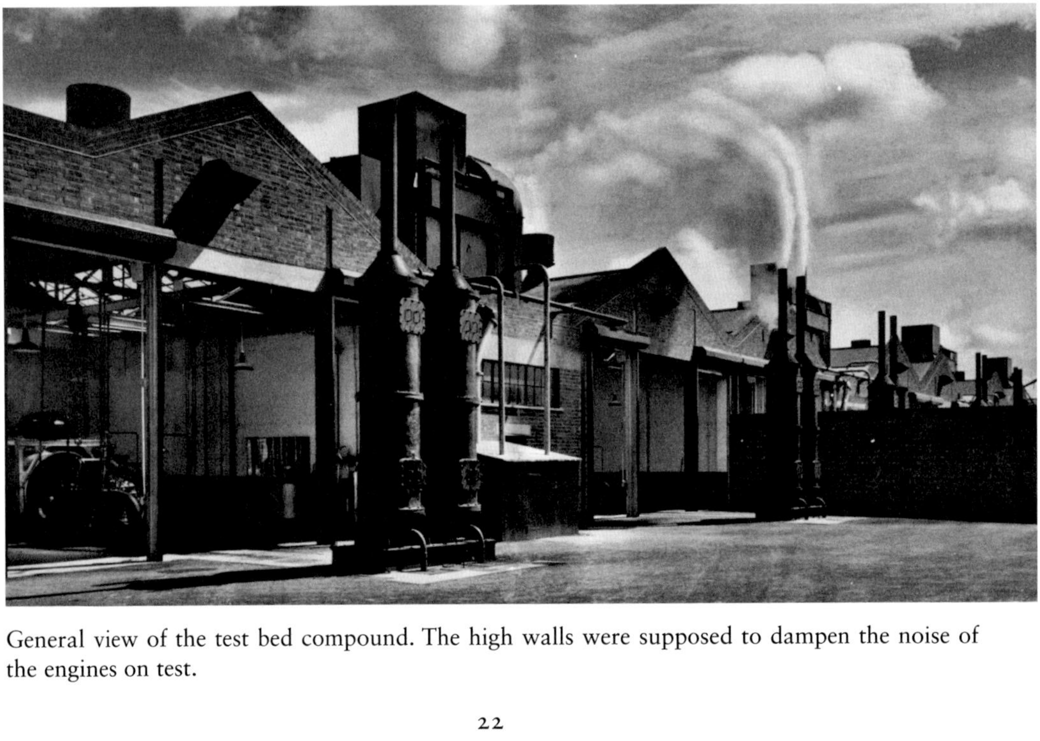

General view of the test bed compound. The high walls were supposed to dampen the noise of the engines on test.

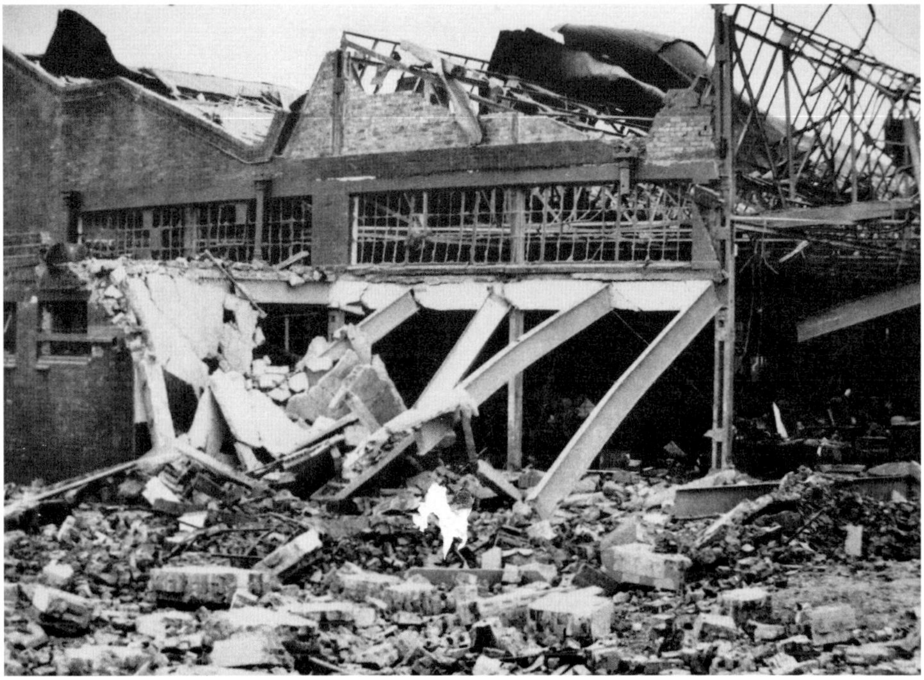

The greatest single tragedy involving loss of life in war-time Crewe was the bombing of Rolls-Royce in December 1940. Two bombs were released from a low flying Junkers 88 which clipped the western extremity of the second machine shop. Production was unaffected.

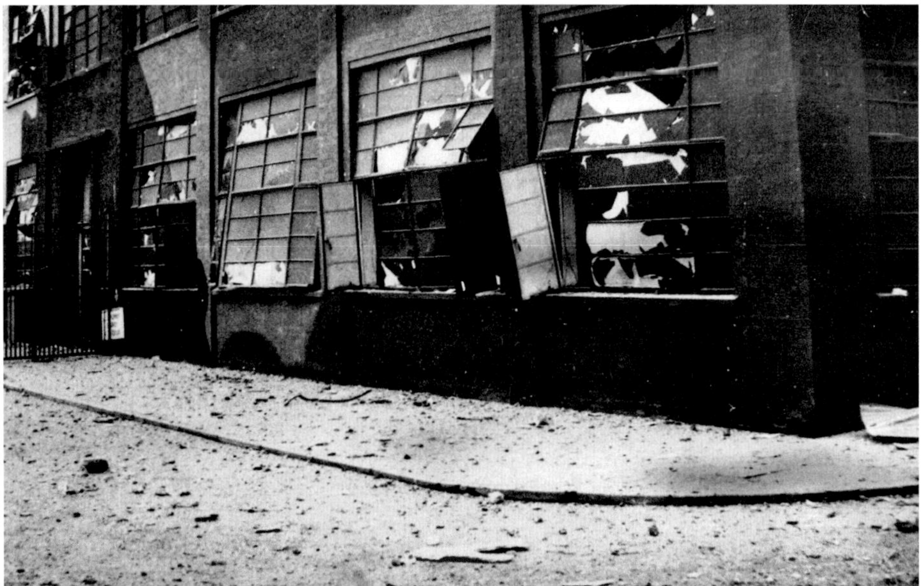

December 1940 looking north. Collateral damage caused by blast from the bombs. Behind the short fence is Pym's Lane and the door of the employment office.

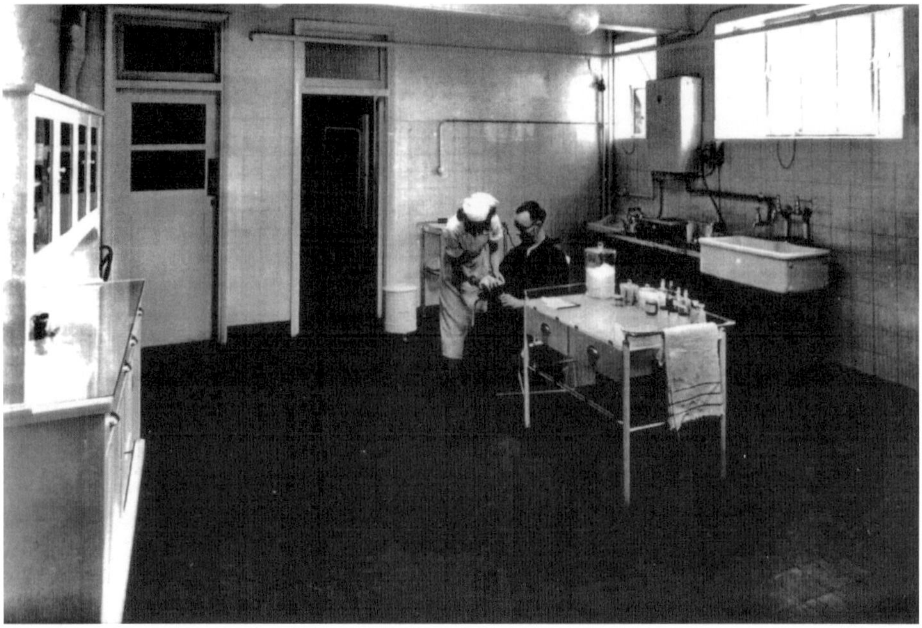

The ambulance room on the east side of the main shop. It had separate entrances for male and female.

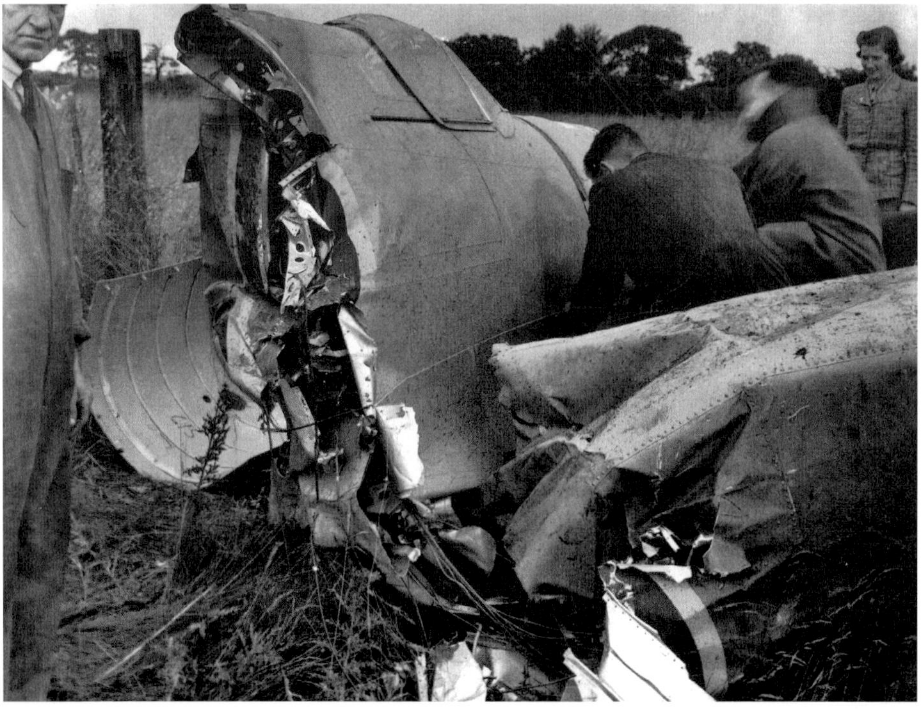

Engine failure on Spitfire EP615 caused this crash at Willenhall in July 1942. Many books, including Alex Henshaw's *Sigh for a Merlin* report the Merlin skew-gear failures yet none trace them to Pym's Lane which is where this author is convinced the fault originated.

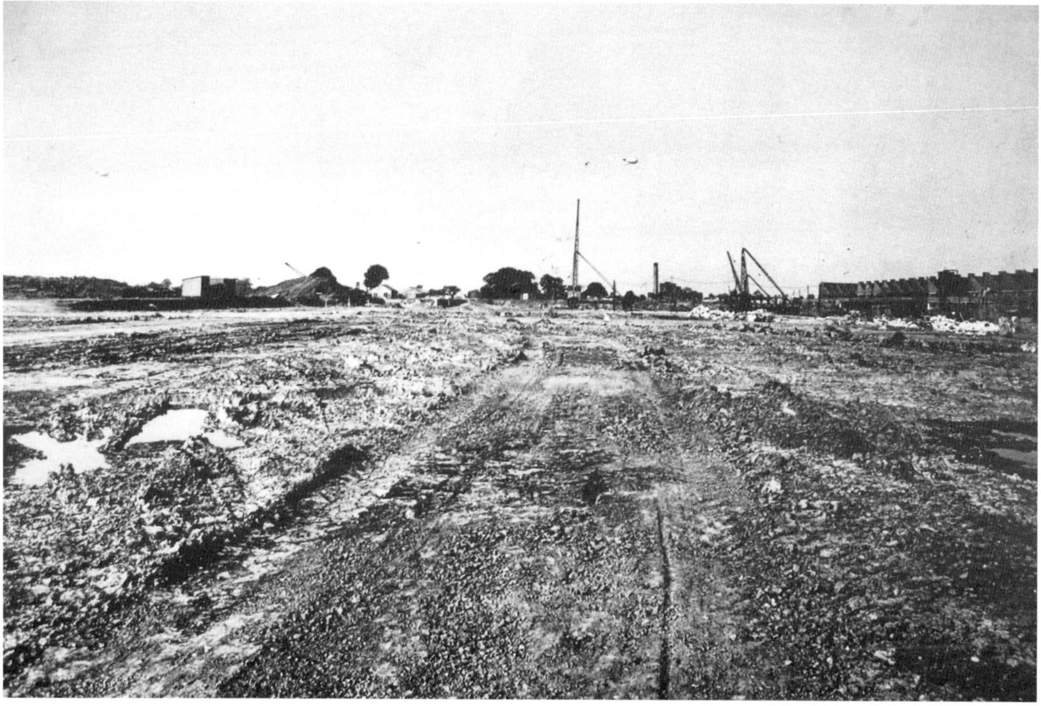

Pym's Lane site looking north 1942 when building commenced on a new workshop for manufacturing the Griffon aero engine. Notice the balloon flying and also the heap of clay known to later RR workers as 'Hill 60'.

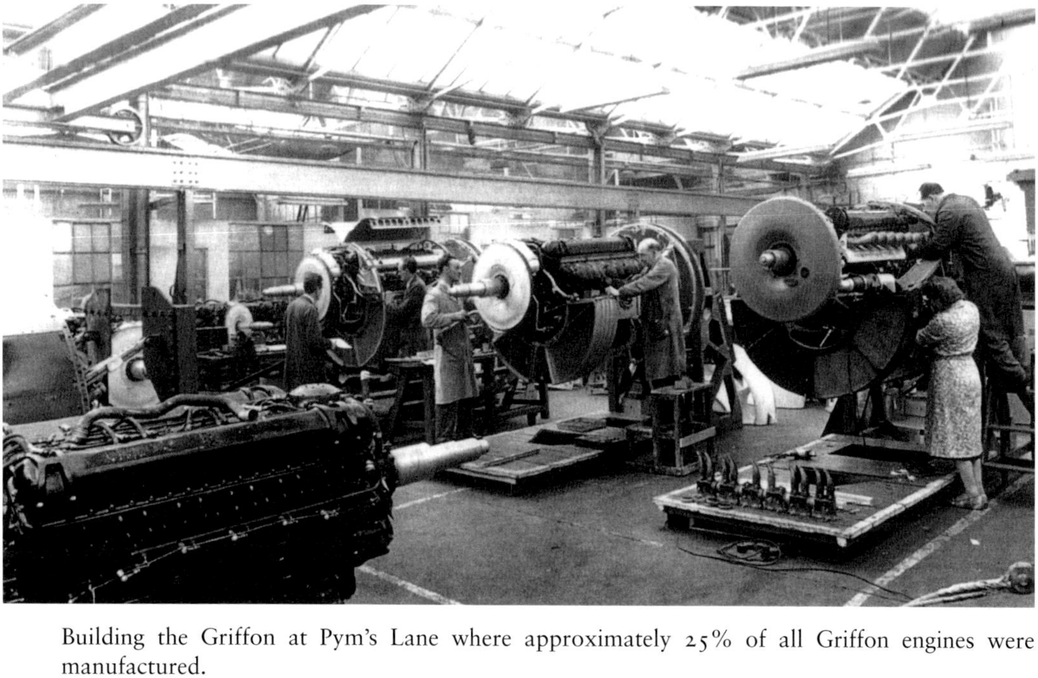

Building the Griffon at Pym's Lane where approximately 25% of all Griffon engines were manufactured.

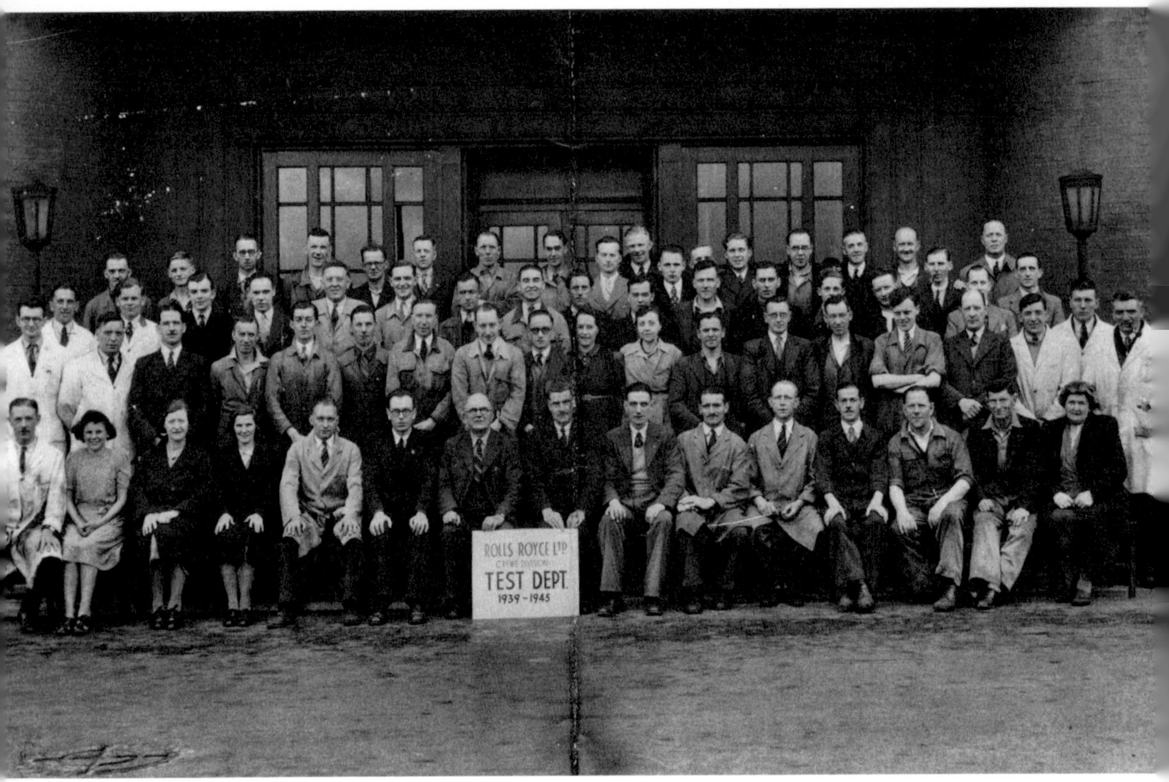

The engine test department in 1945. Wilf Thomas who assembled the differential mechanism in the forties and fifties and known by many engineering apprentices for his educated conversation is third from the left on the back row.

three

A NEW DAWN

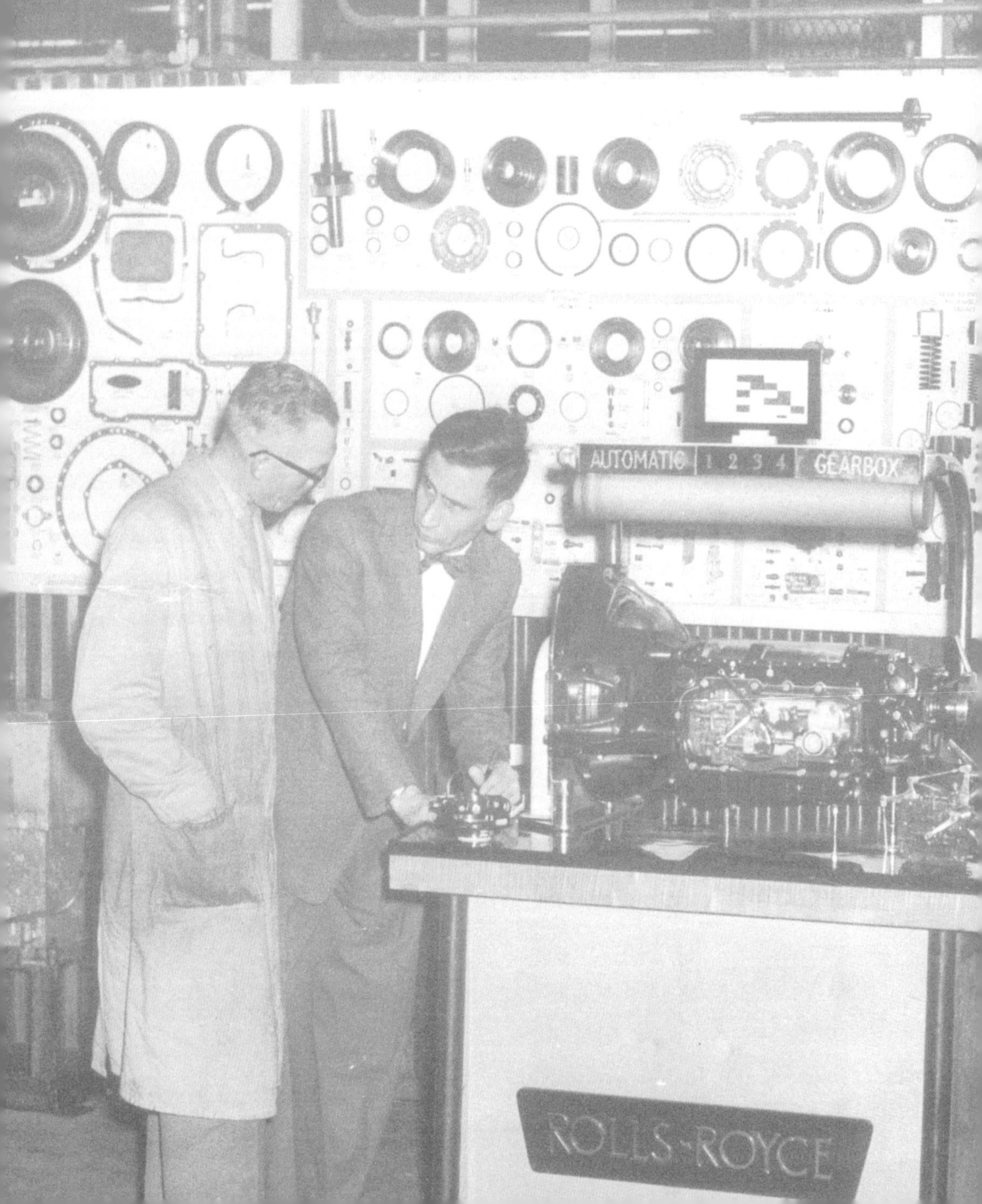

In 1944 Dr Llewellyn Smith was appointed as general manager at Crewe with the brief to plan for car production when the war ended. He and his team overcame many inherent problems finishing a Bentley Mk VI, the first complete car ever to be built by RR, in February 1946. Previously Rolls-Royce had manufactured the chassis and running gear leaving the customer to appoint a specialist coach-builder to body and trim the car though a RR Silver Wraith was still available to be finished this way. In 1949 the Rolls-Royce Silver Dawn (export only) was launched to take its place alongside the Bentley R series. This was also the era when the Bentley Continental was manufactured.

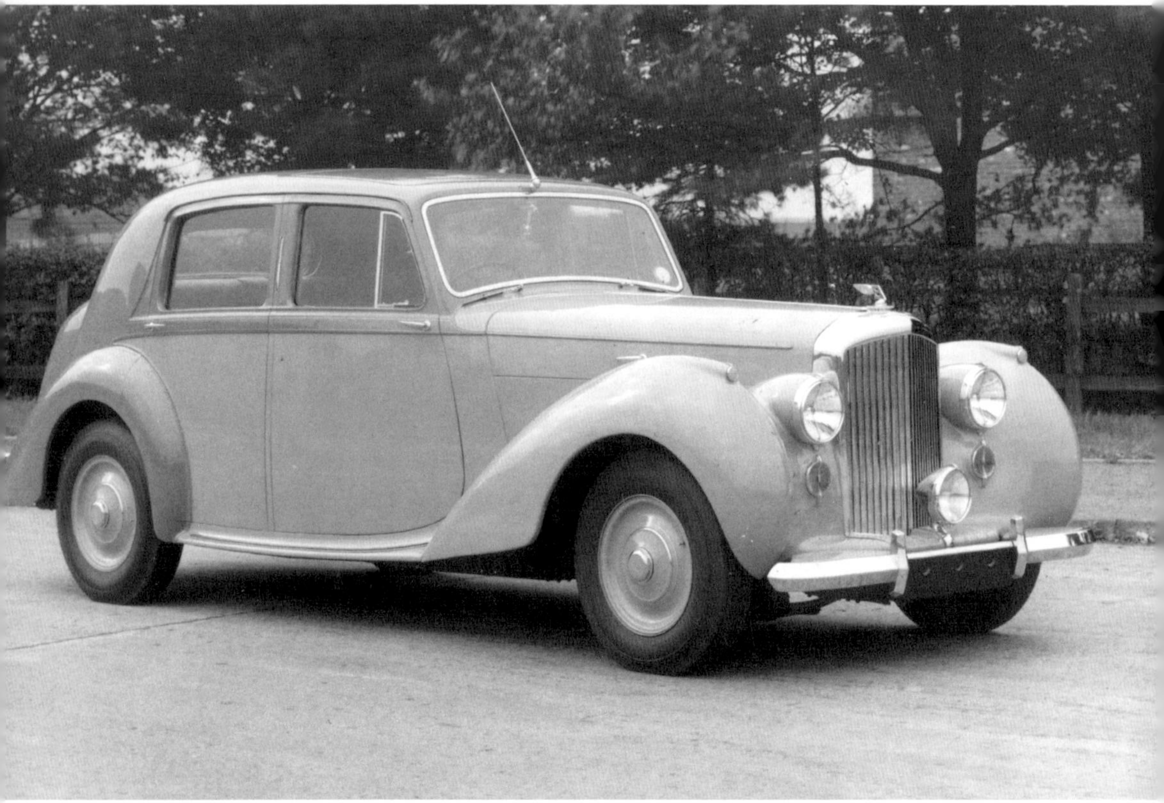

A Bentley Mk VI with a Pressed Steel standard body, the first car to be manufactured at Pym's Lane. Launched in 1946 with a price of £3,317 nearly 5000 were manufactured before being phased out in 1952.

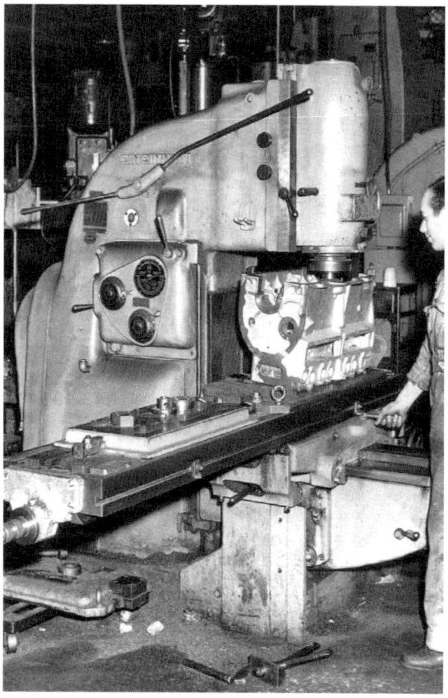

Right: Machining a six cylinder B range crankcase circa 1950. Notice the bolts and other shop tackle littering the floor.

Below: Operating a Ward 2A capstan lathe. In the spring of 1939 there was a strike over women working such machine tools.

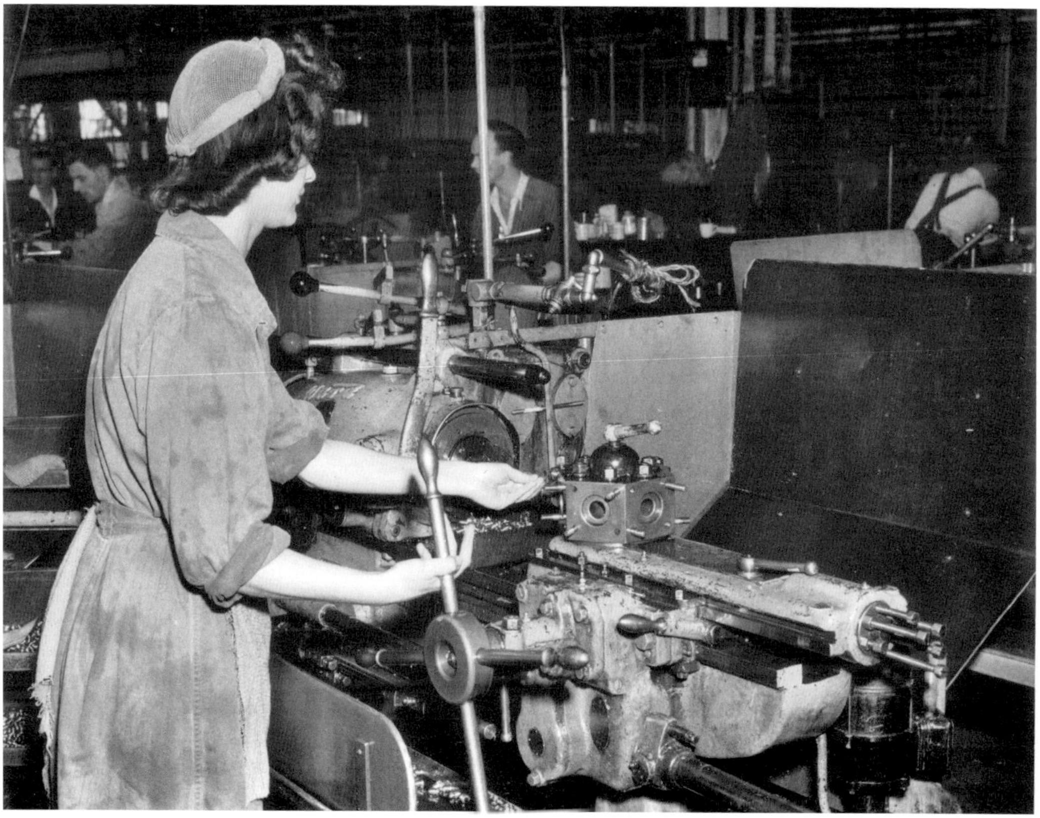

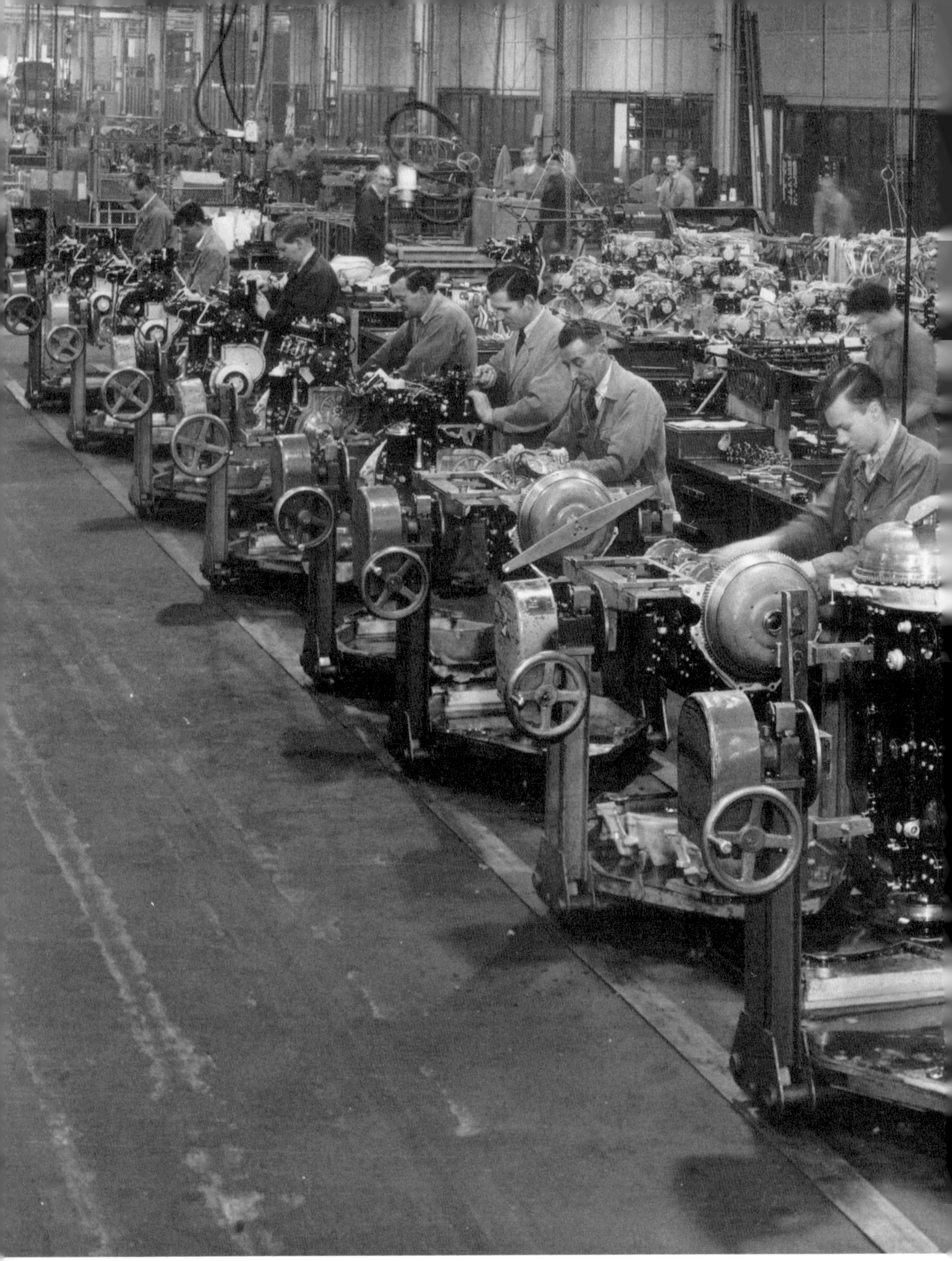

Engine build (152 department) looking south towards the chassis line. The windows along the gangway belong to the progress office.

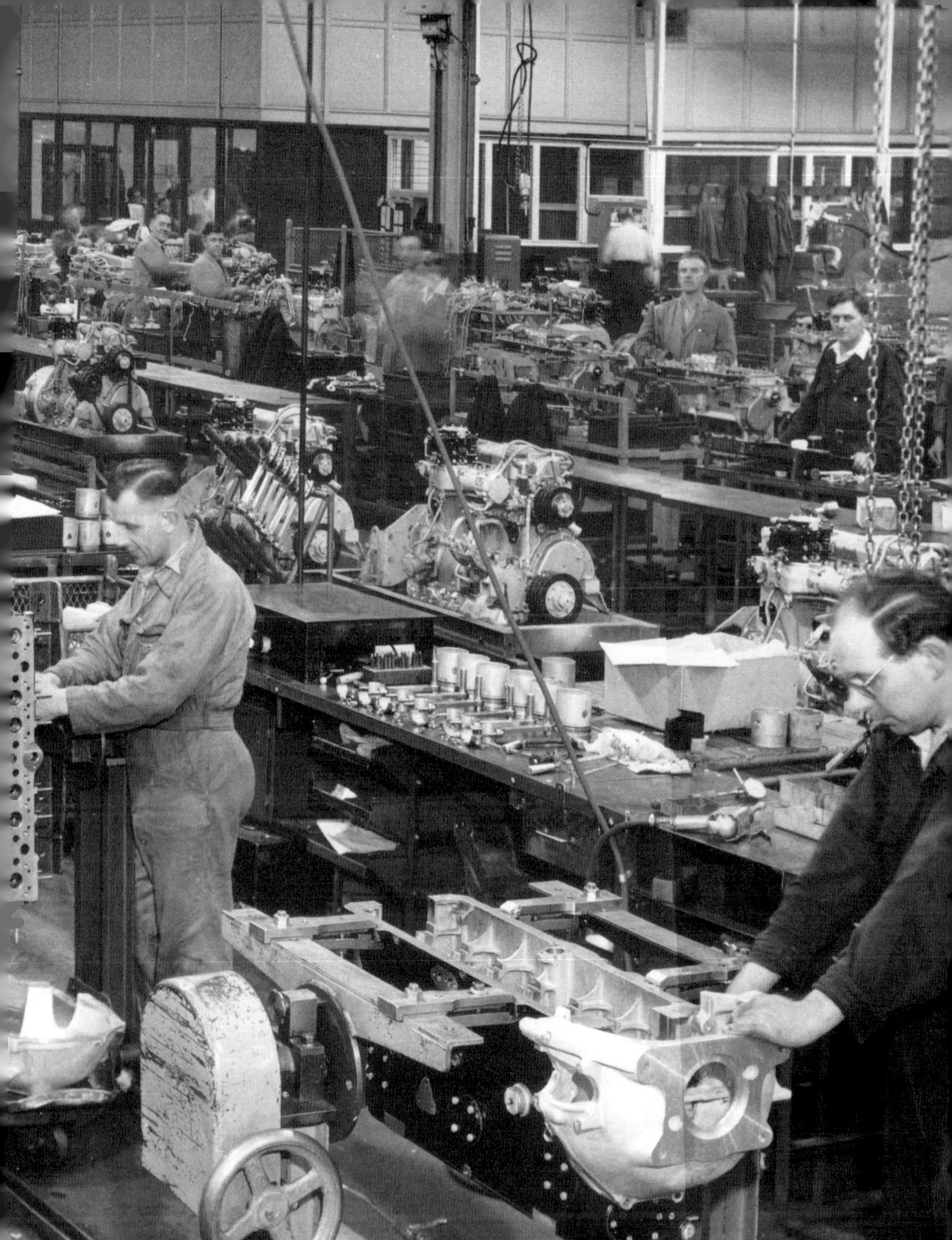

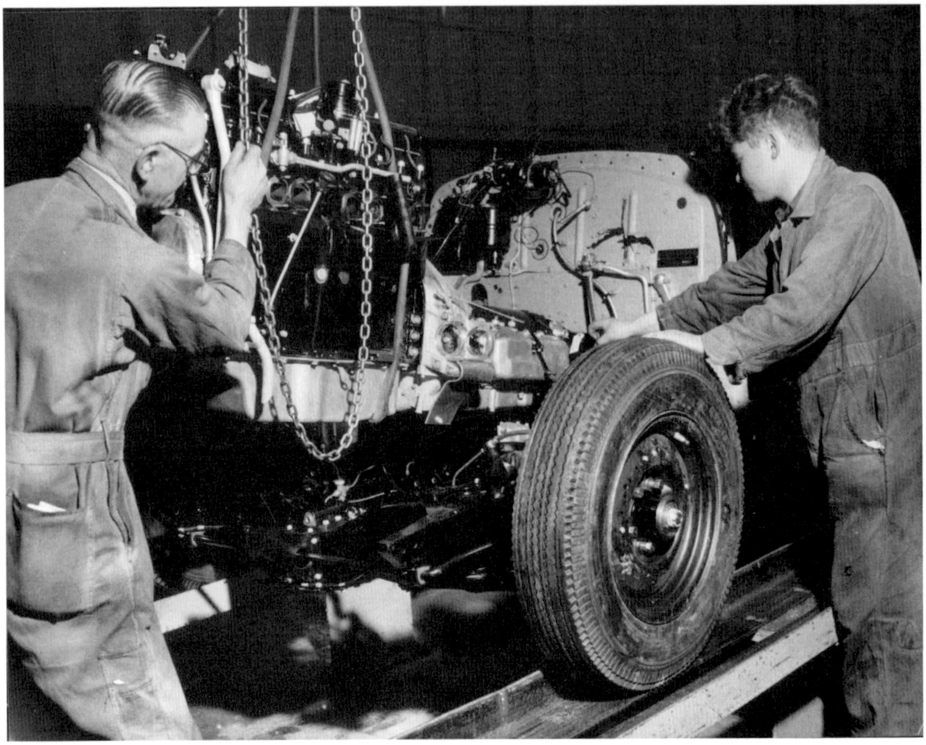

Fitting the six cylinder engine on the chassis line, a job for a fitter and an apprentice.

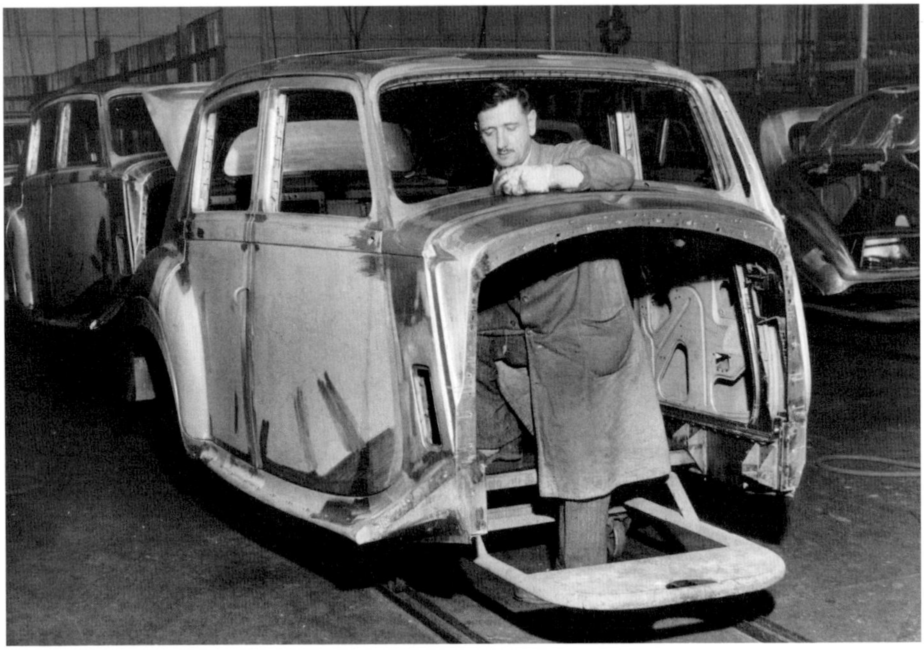

Inspection of a degreased body after being received from Pressed Steel. At this stage it could be for a Bentley or a Silver Dawn.

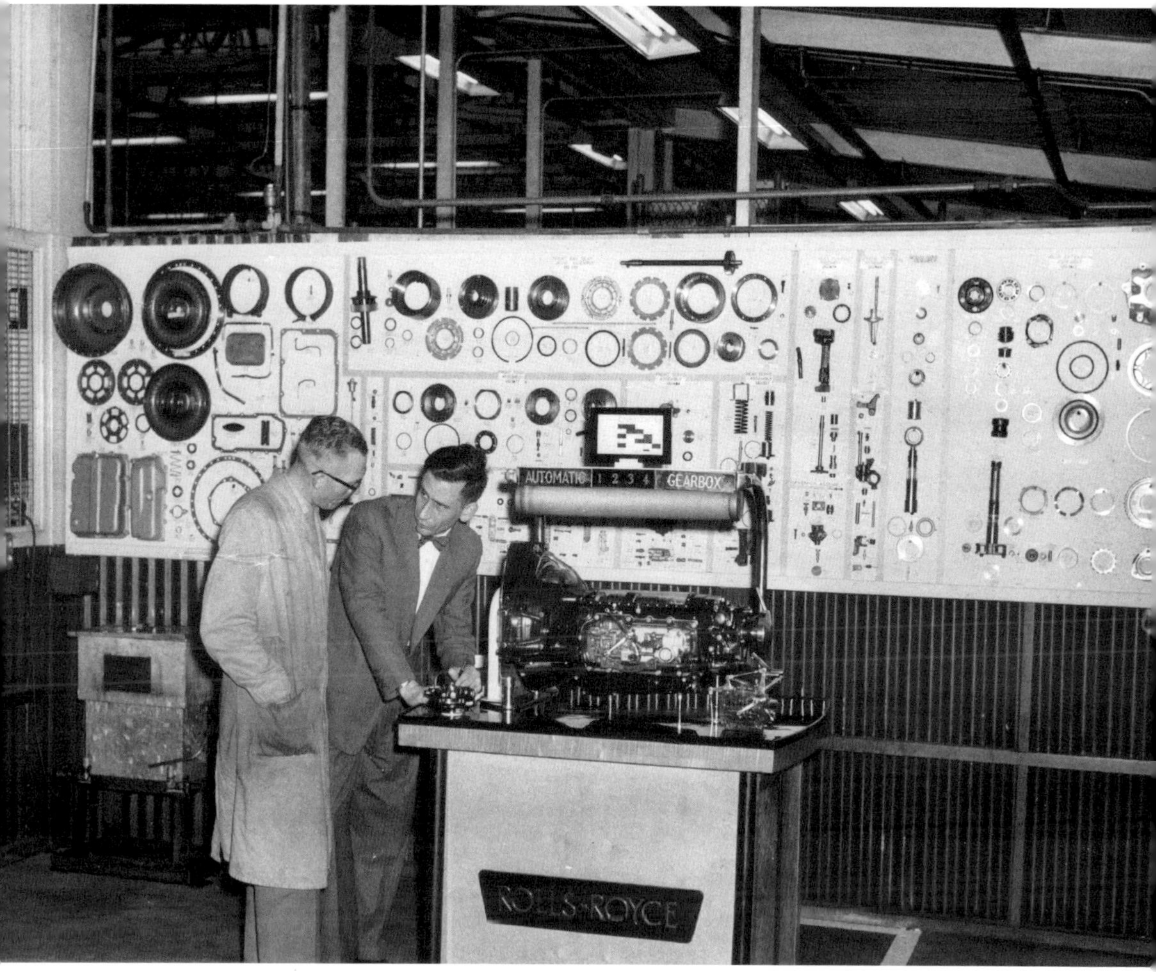

In 1952 an automatic gearbox was available for export only models. This one has been sectioned to allow David Scott to explain the inner workings to visitors. This, or a similar model, is now at Paulerspury.

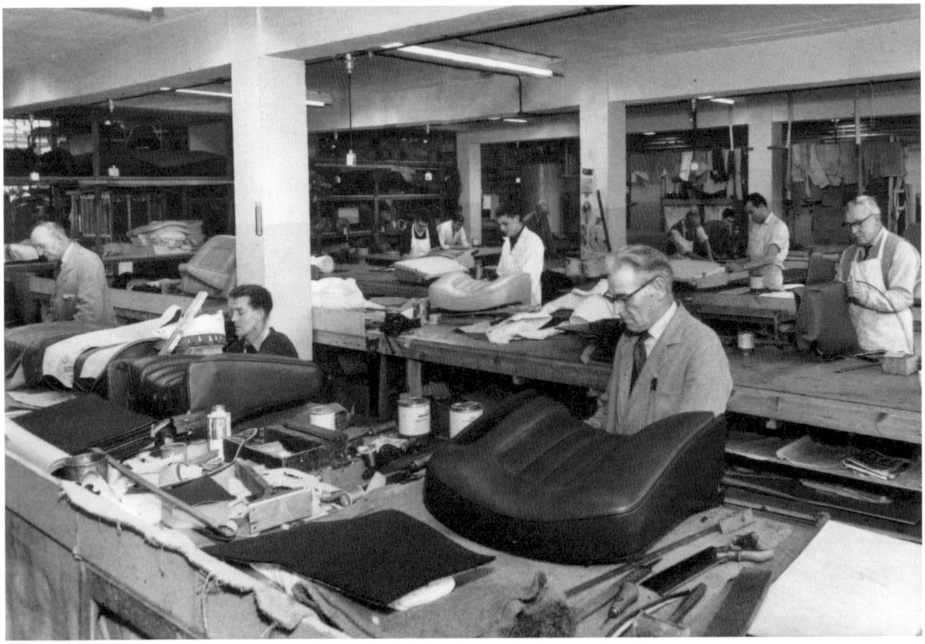

General view of the upholstery department. In the early post-war years it was not easy to attract craftsmen experienced in coach-trimming techniques.

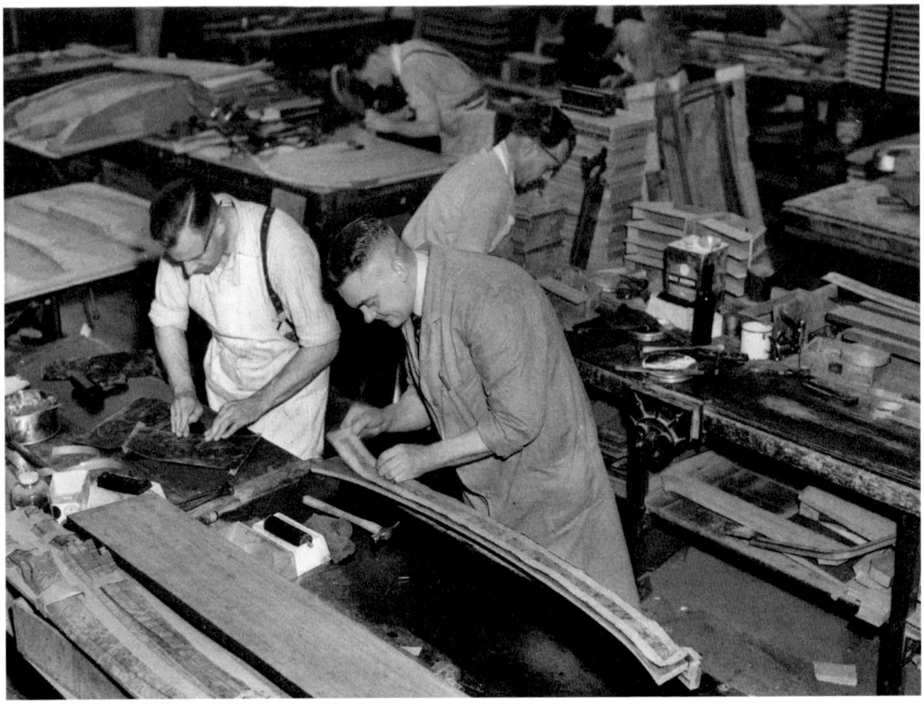

A view of the cramped working conditions in the wood-shop. Until RR was able to expand into the buildings in the west of the site, space was always at a premium.

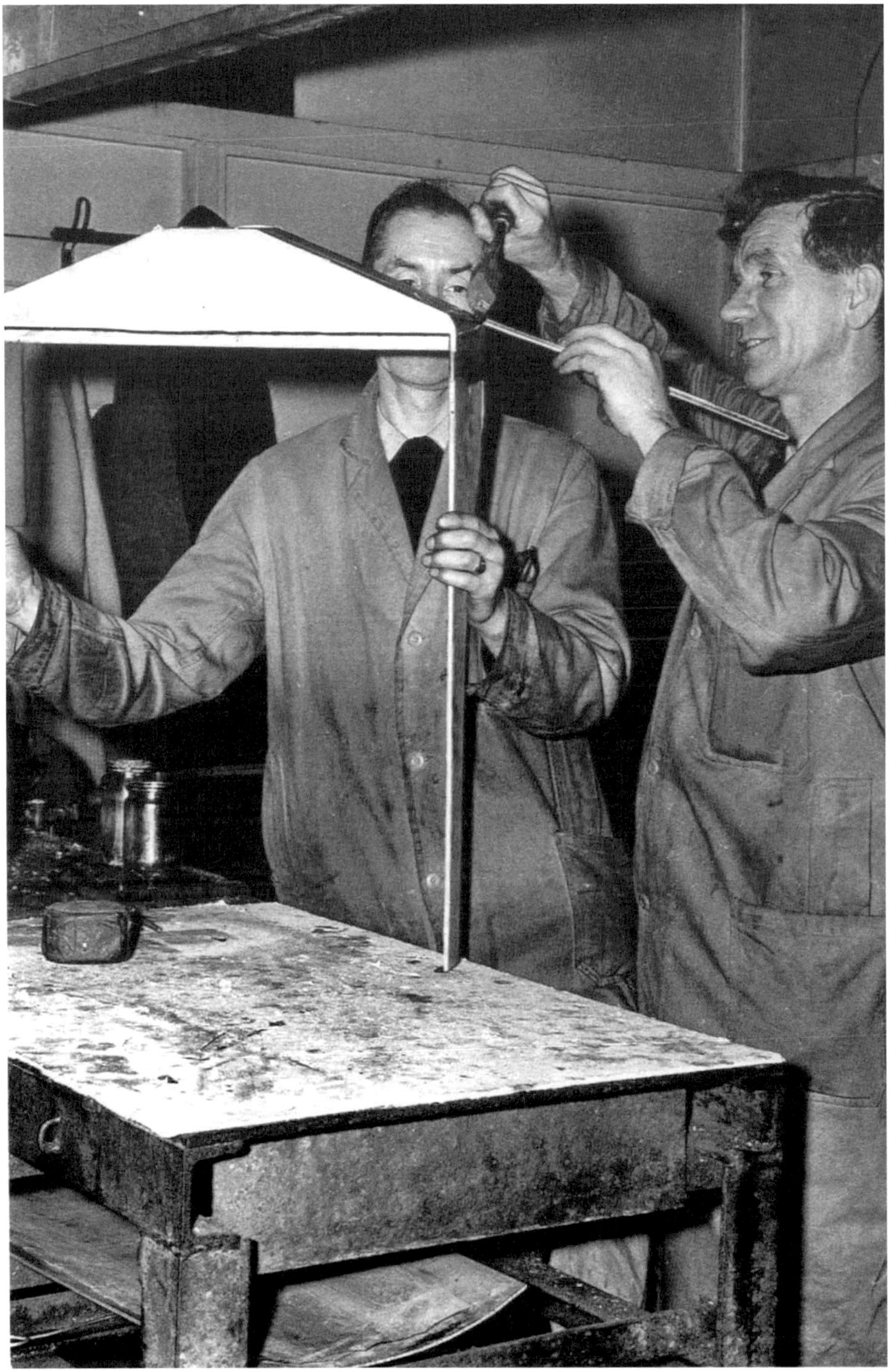

A Rolls-Royce radiator shell being manufactured by hand. When completed the craftsman always etched his initials on the inside.

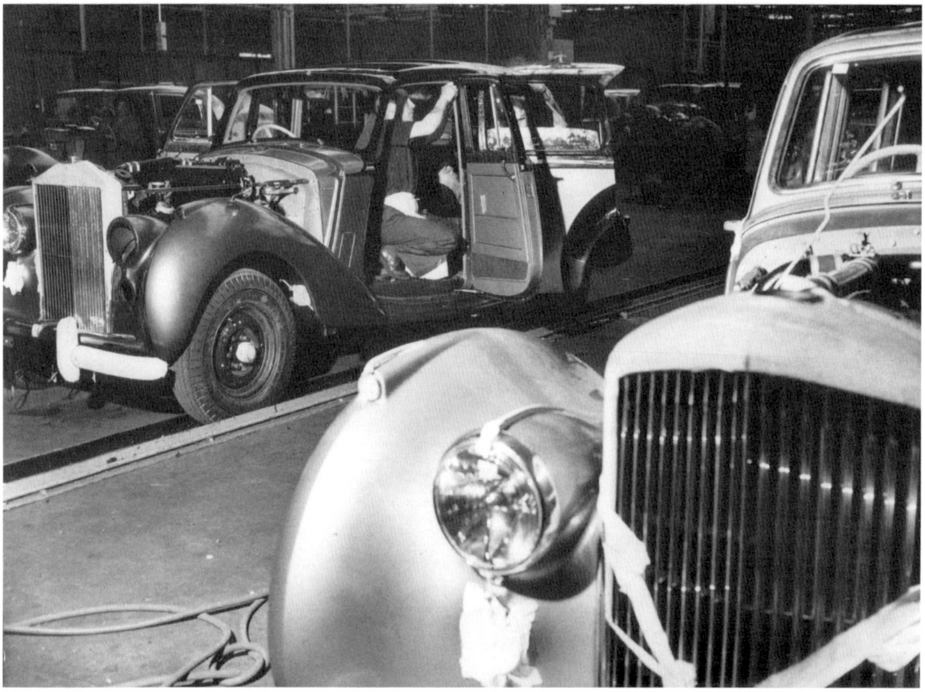

A Silver Dawn and an R type Bentley being trimmed on the body line at the south end of the main shop. In 1953 a Silver Dawn cost £4,605.

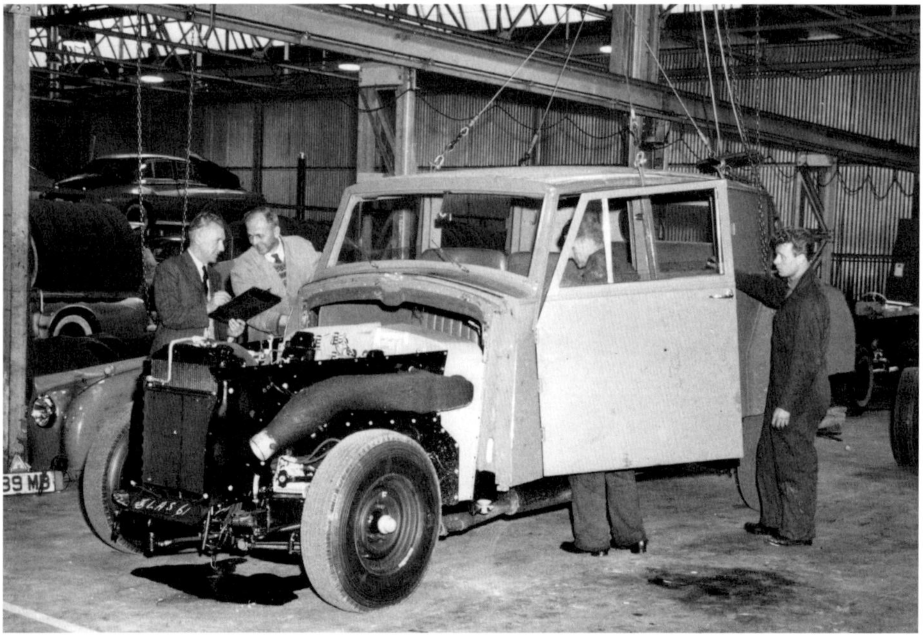

Doug Fox, chief tester supervising the mounting of a 'slave body' on to a Silver Wraith chassis prior to road testing.

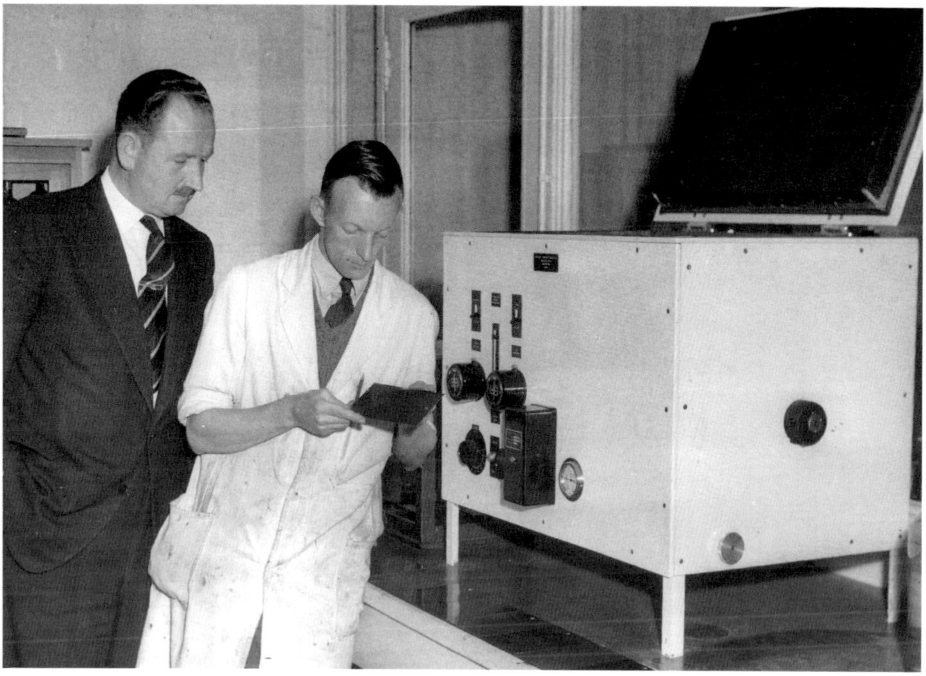

Examining a paint sample in the laboratory. The man in the suit is Basil Wilmin, the chief paint technician, and one of the best dressed men at Crewe RR.

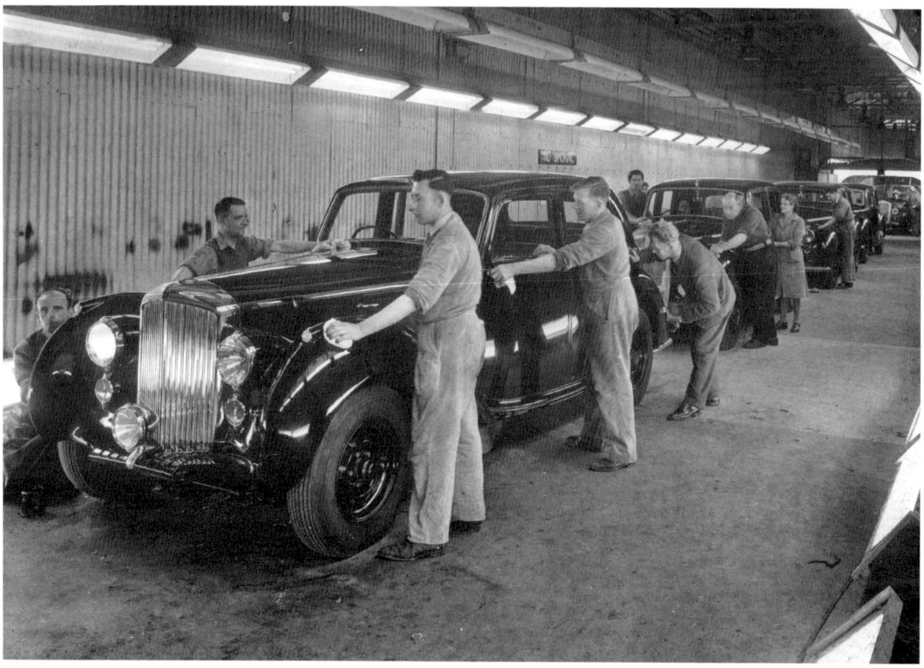

After leaving the paint shop a Bentley receives its final polish in the area of the factory that was bombed in 1940. The posture of the polishers suggests they were posing.

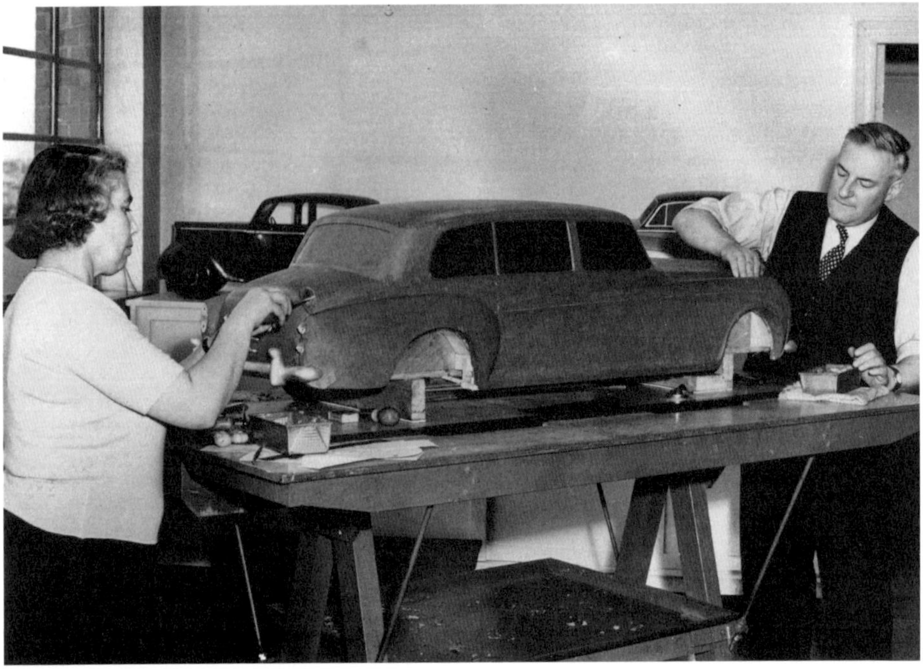

Miss C. Jenner and Bill Allen working on a quarter scale model of a RR Phantom V.

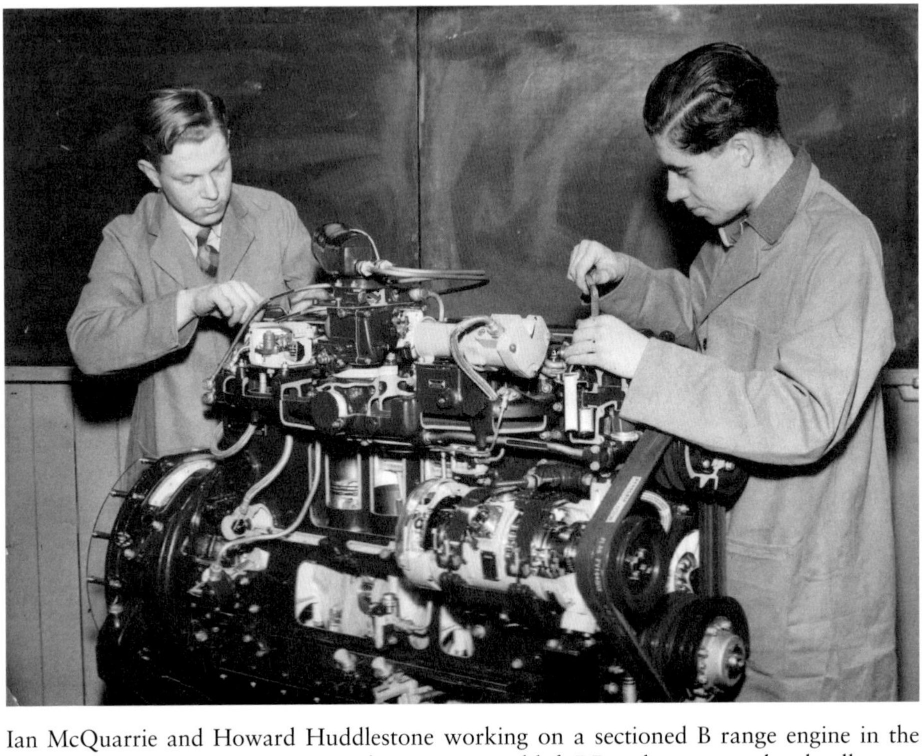

Ian McQuarrie and Howard Huddlestone working on a sectioned B range engine in the apprentices' school in the autumn of 1954. Howard left RR to lecture at a local college.

four

SILVER CLOUDS WITH BLUE STREAKS

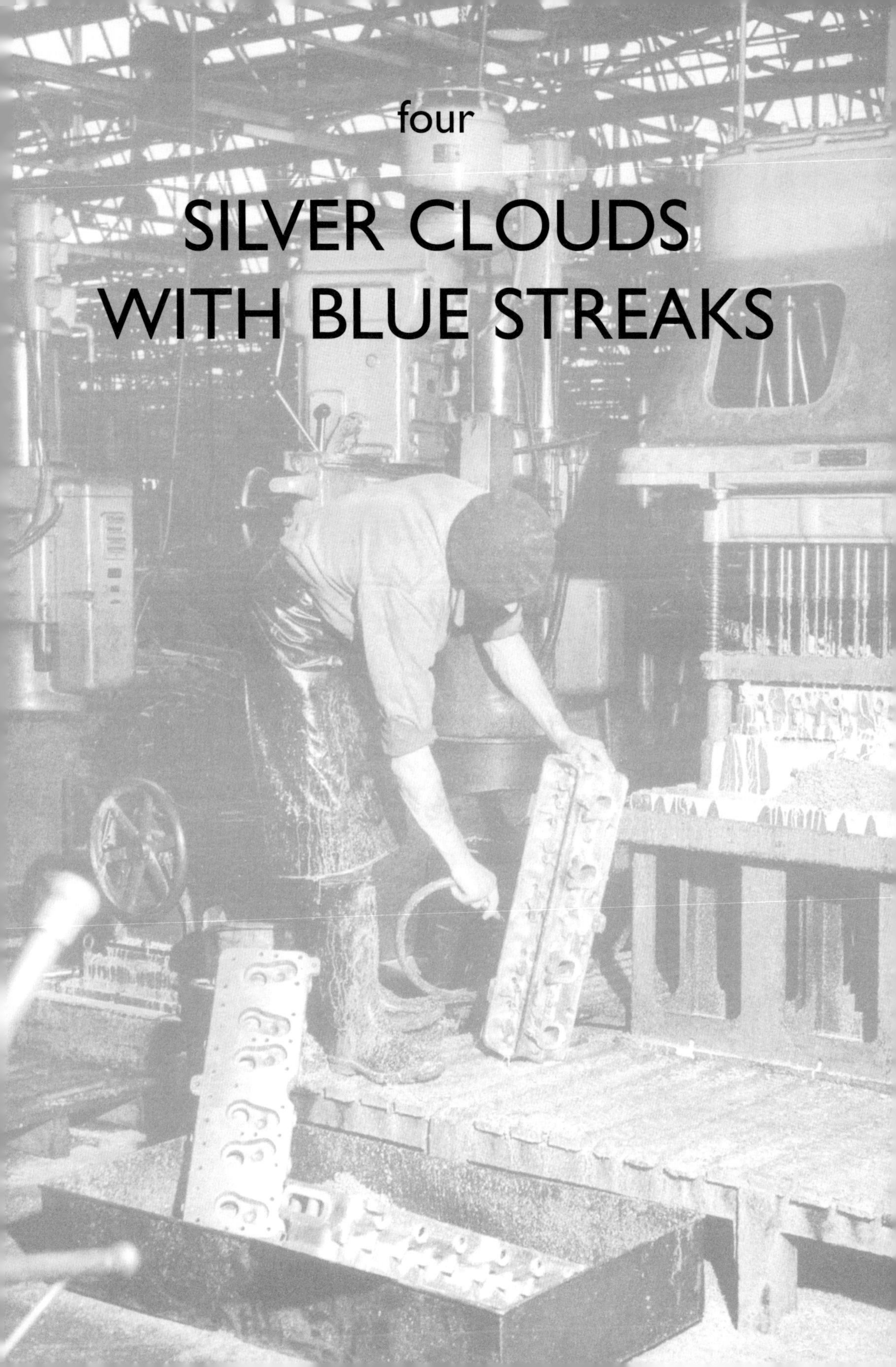

The major change to the cars during this era was the replacement of the six cylinder engine first introduced in the 1920s. By the early 1950s it had reached the end of its development meaning that Charlie Jenner and later Jack Phillips were charged with designing another. So successful were they that the V8 engine, installed into the Silver Cloud and Bentley S series in 1959, lasted, (much developed) for the rest of the twentieth century. It should be noted that the RR Silver Cloud and Bentley S series, launched in 1955, were phased out in 1965.

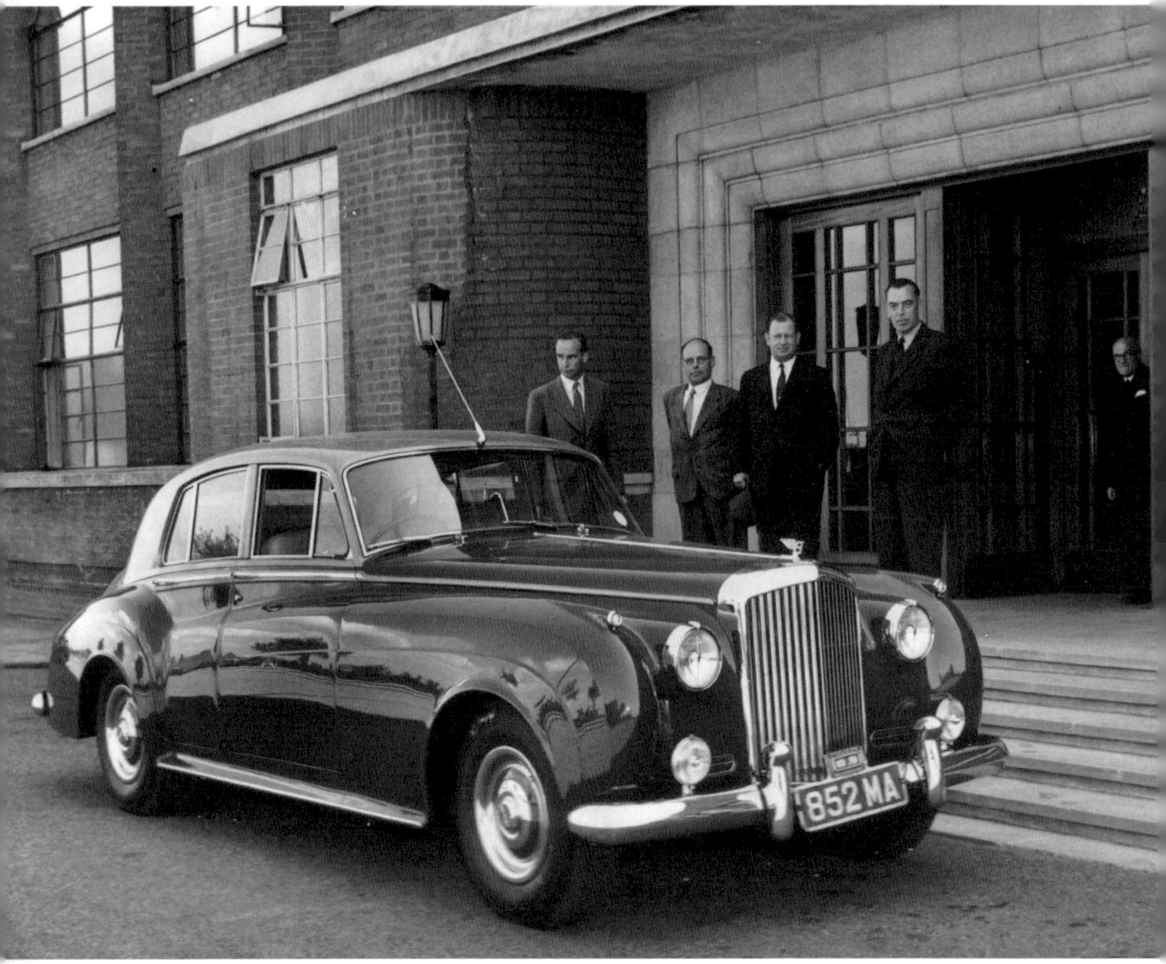

Bentley S series outside of the main entrance. Fred Stafford, the works' manager is standing on the top step. In 1955 it was priced at £4,670.

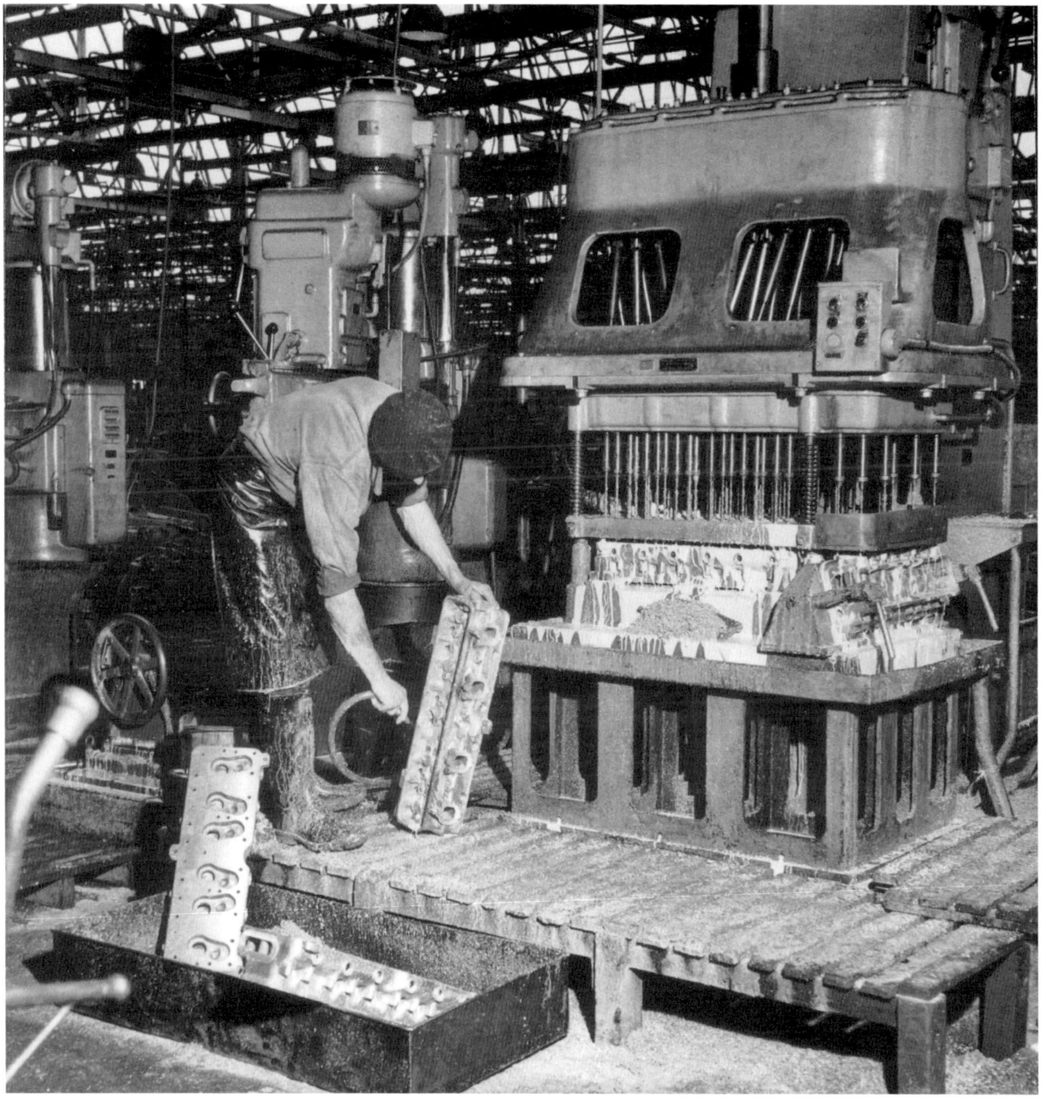

A multi-spindle machine drilling a cylinder head, an operation that generated a lot of metal chips or 'swarf'. Notice the coolant or 'bosh' flooding over the drills. This gave the main shop its distinctive smell. Notice also the wellington boots worn by the operator to keep his feet free from coolant and swarf.

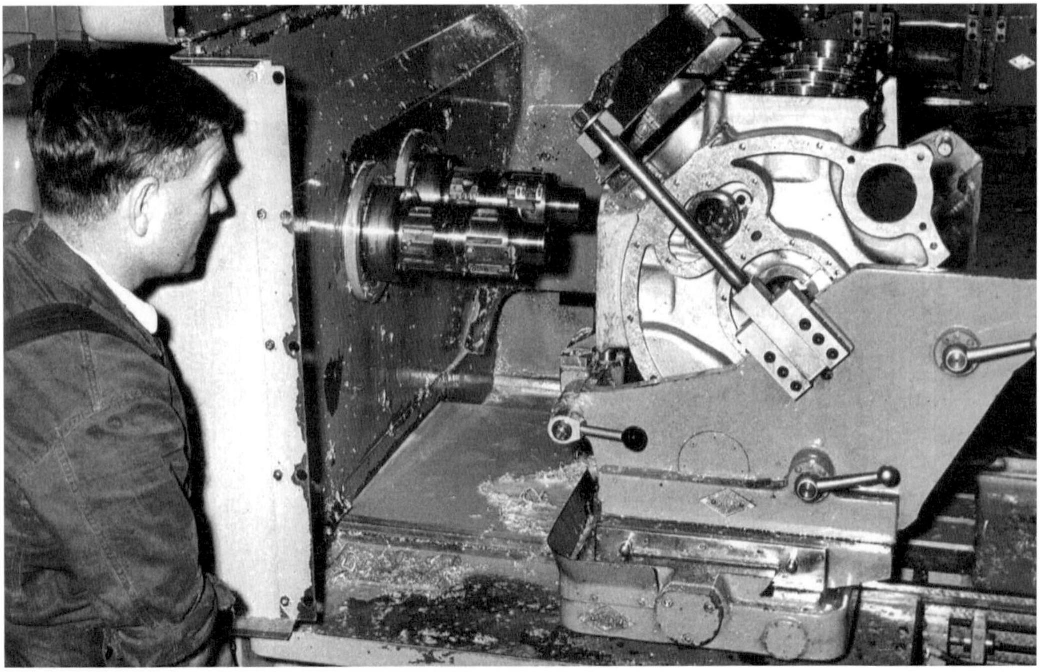

Boring a V8 crank-case. The skill in this job lay with the tradesmen who set the machine.

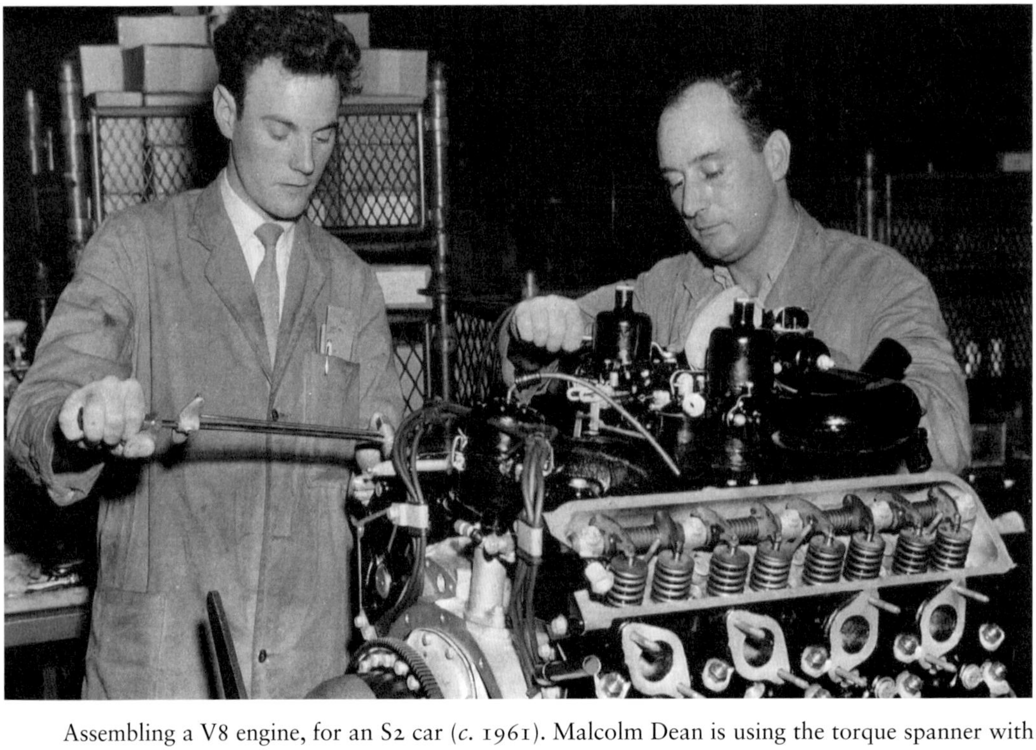

Assembling a V8 engine, for an S2 car (*c.* 1961). Malcolm Dean is using the torque spanner with Tom McGowan alongside.

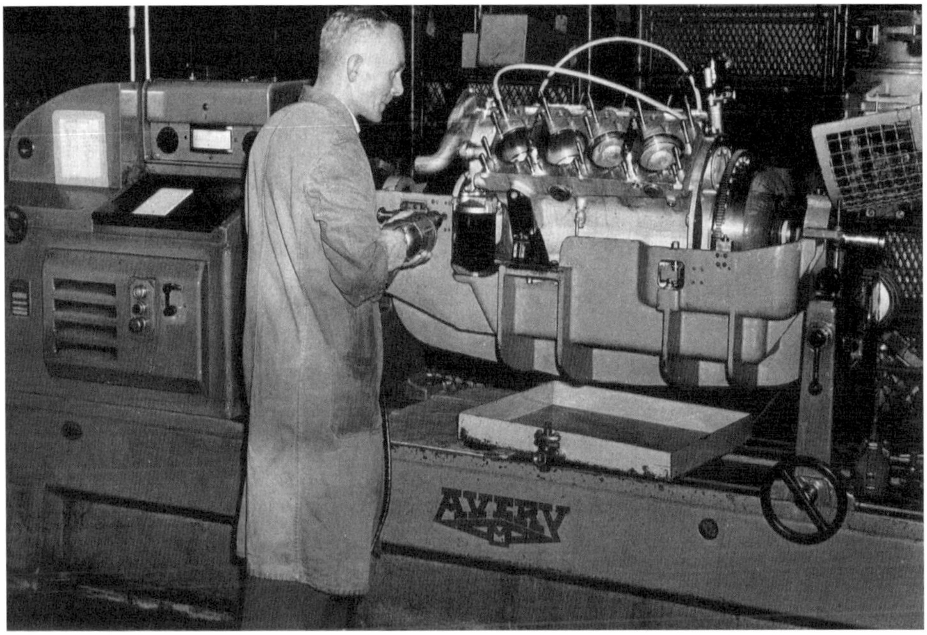

Final balancing check of the engine when being assembled to ensure smooth running.

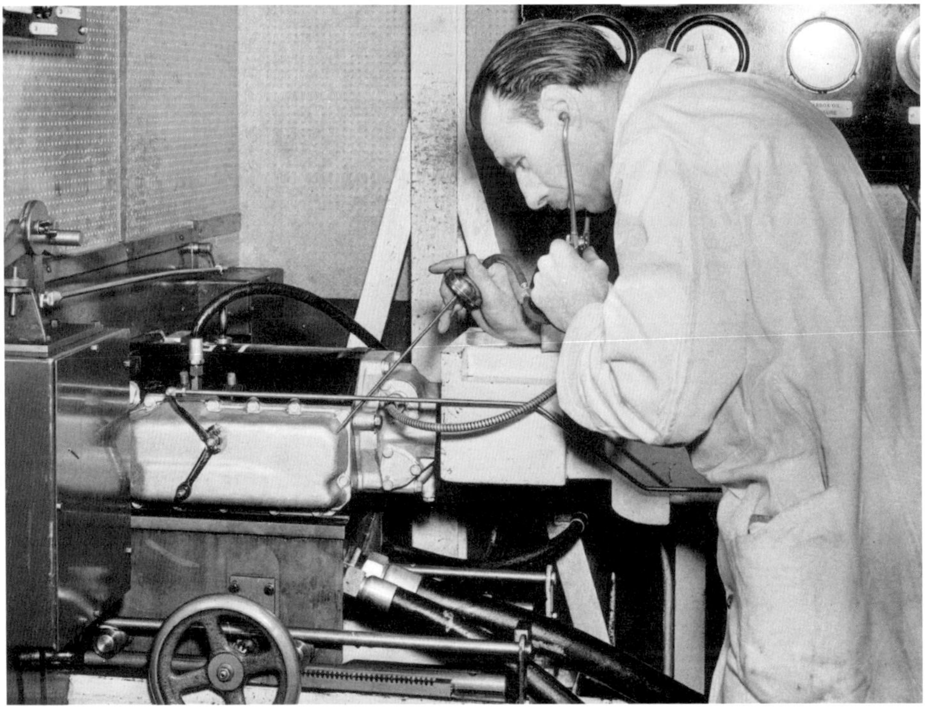

Automatic gearbox being checked on a test rig (*c.* 1956).

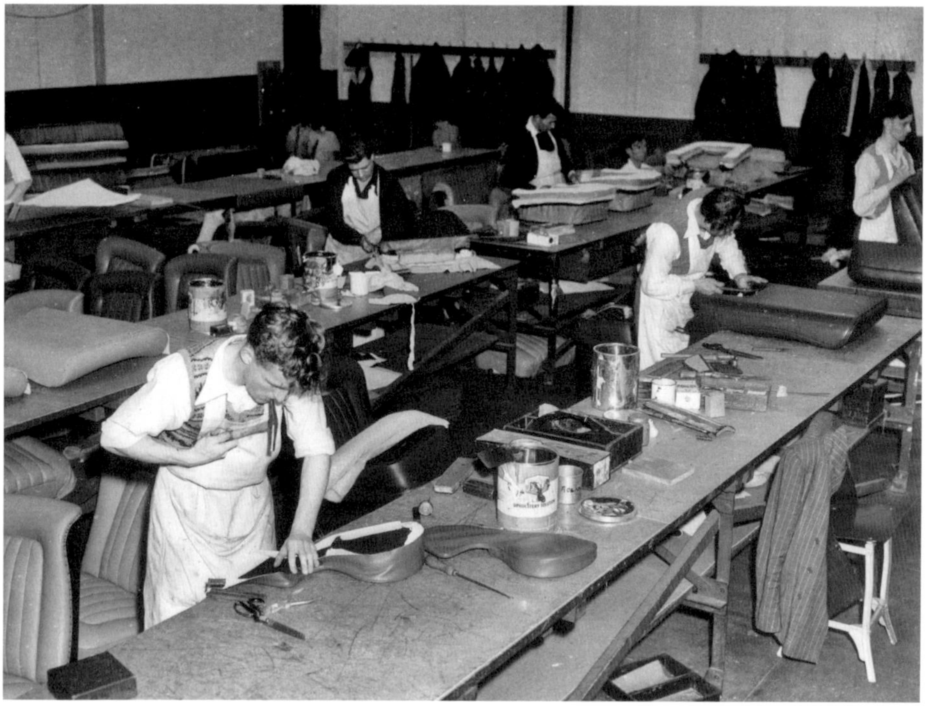

General view of the trim shop (*c.* 1959). Bob Moulton, extreme right, was the first indentured apprentice when the scheme was commenced in the late forties.

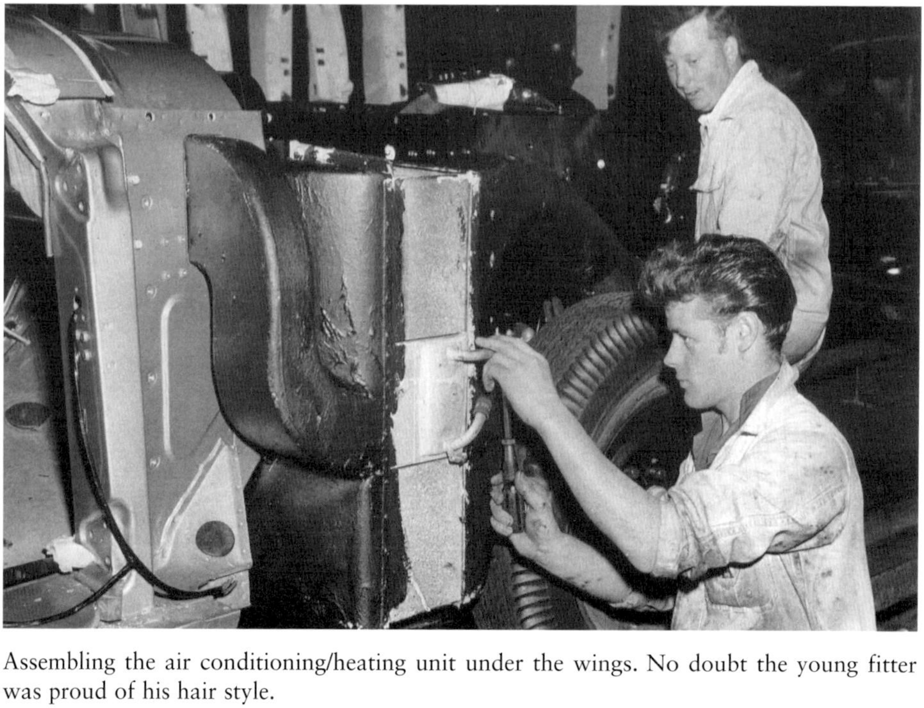

Assembling the air conditioning/heating unit under the wings. No doubt the young fitter was proud of his hair style.

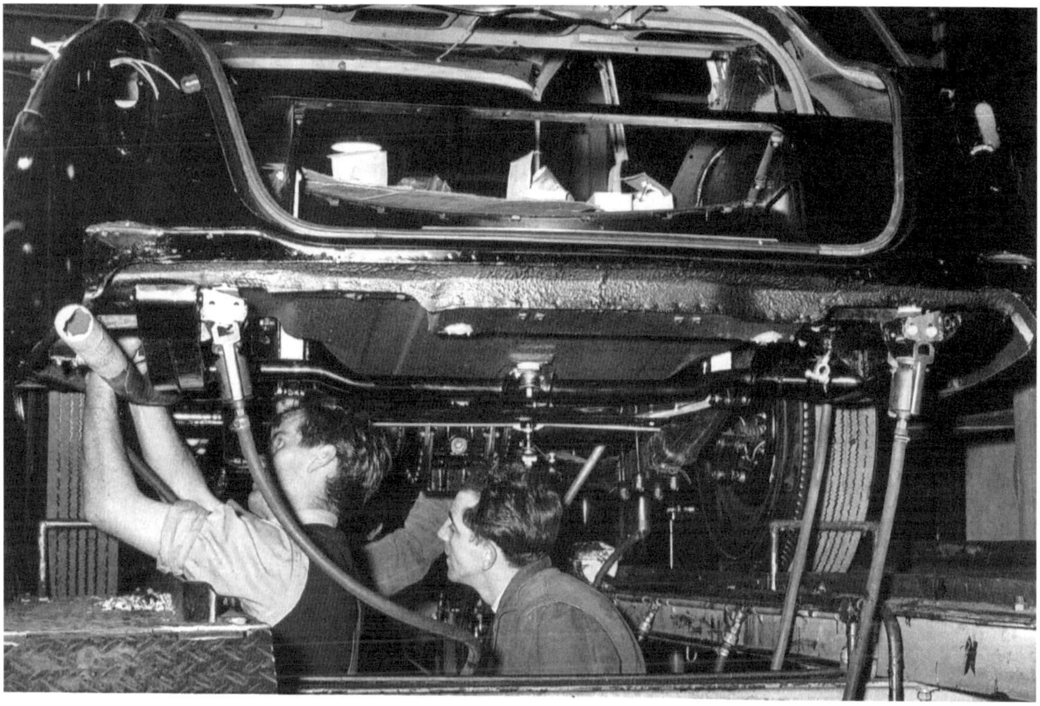

Mounting the body on to the chassis. The special rig ensures each point is sustaining an identical load. Working in this position made an inexperienced fitter's arms' ache.

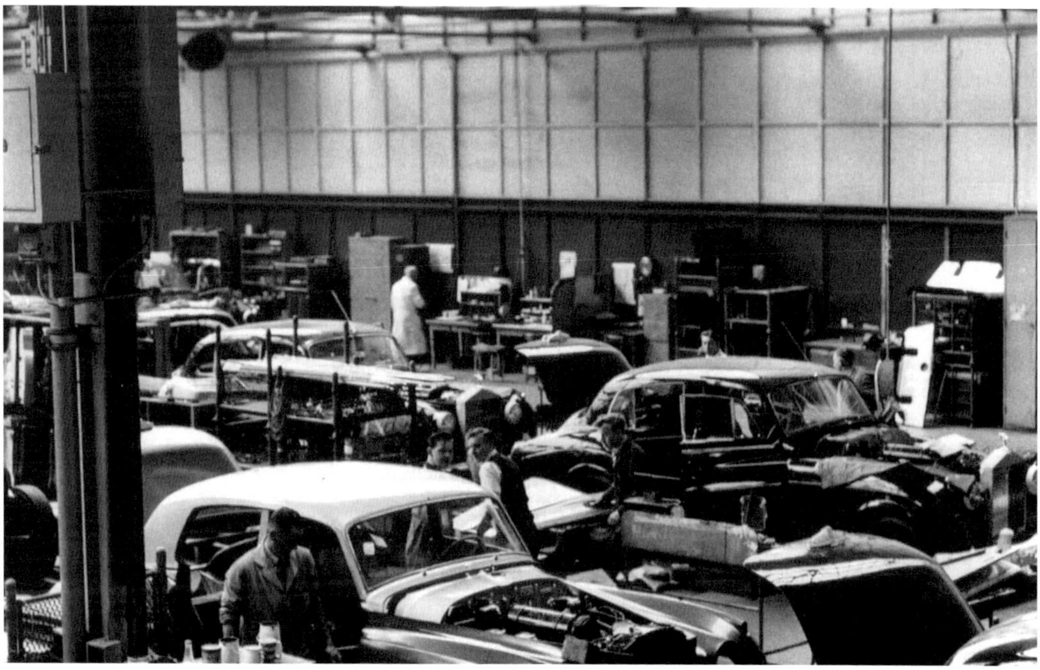

Bentley S series and RR Silver Cloud on the assembly line (c. 1958).

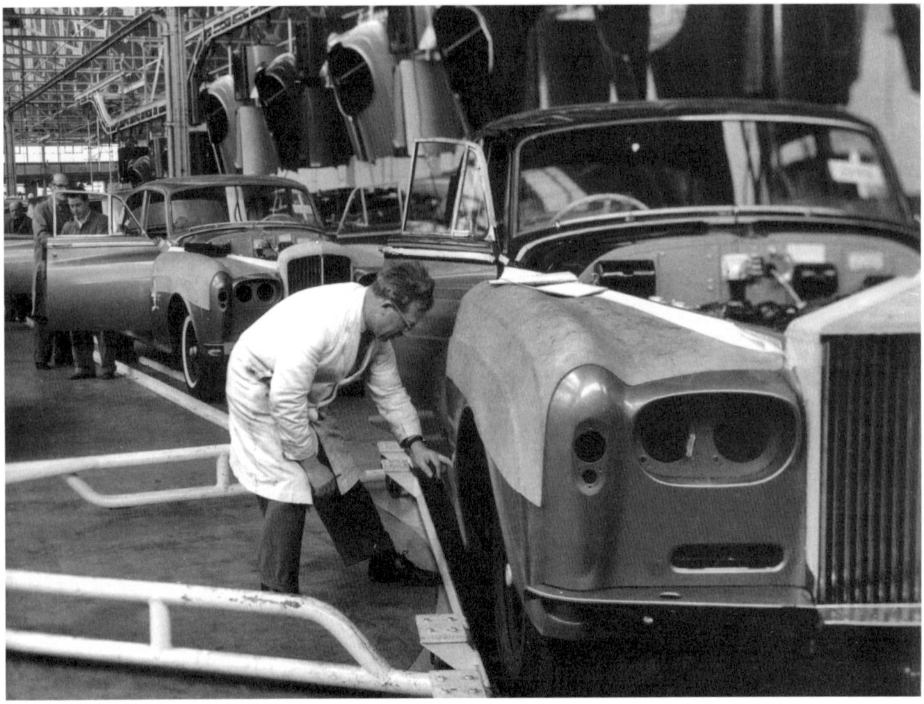

Fitting the front wings to a RR Silver Cloud III (*c.* 1963).

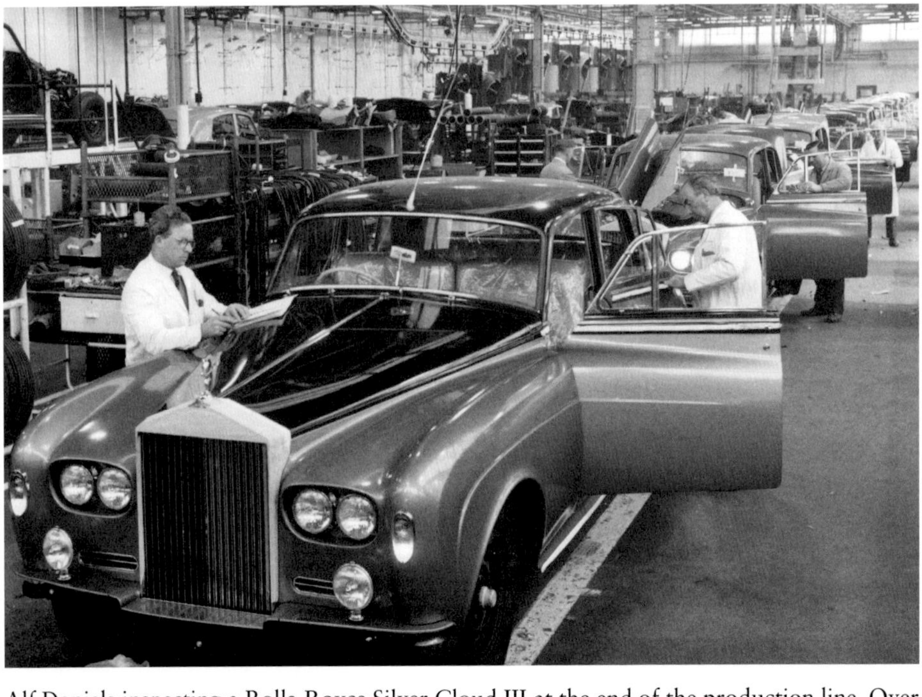

Alf Daniels inspecting a Rolls-Royce Silver Cloud III at the end of the production line. Over 2280 were sold between 1962 and 1965.

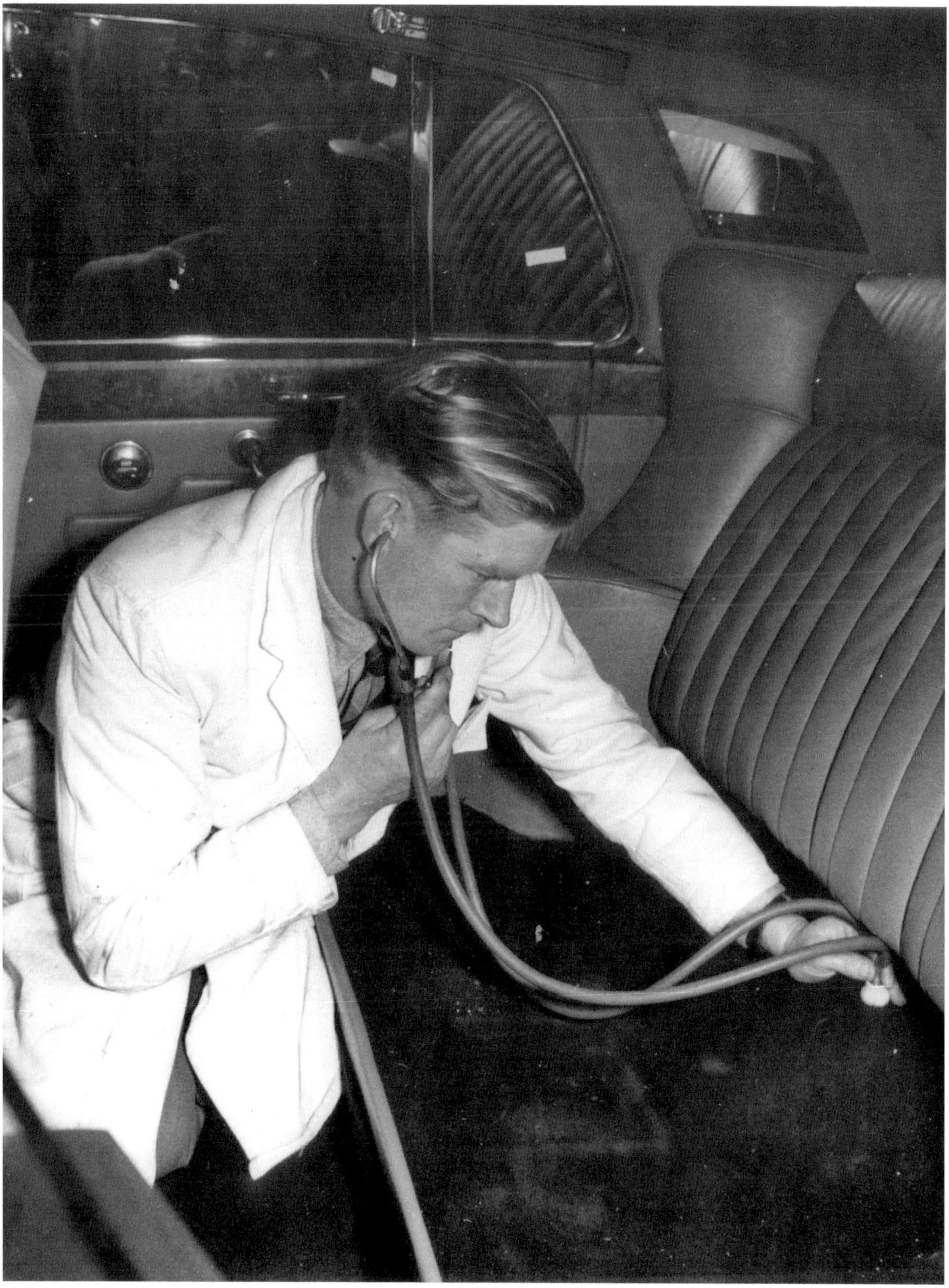

In this posed view Bert Walton is demonstrating how he checked for rear axle whine during road testing.

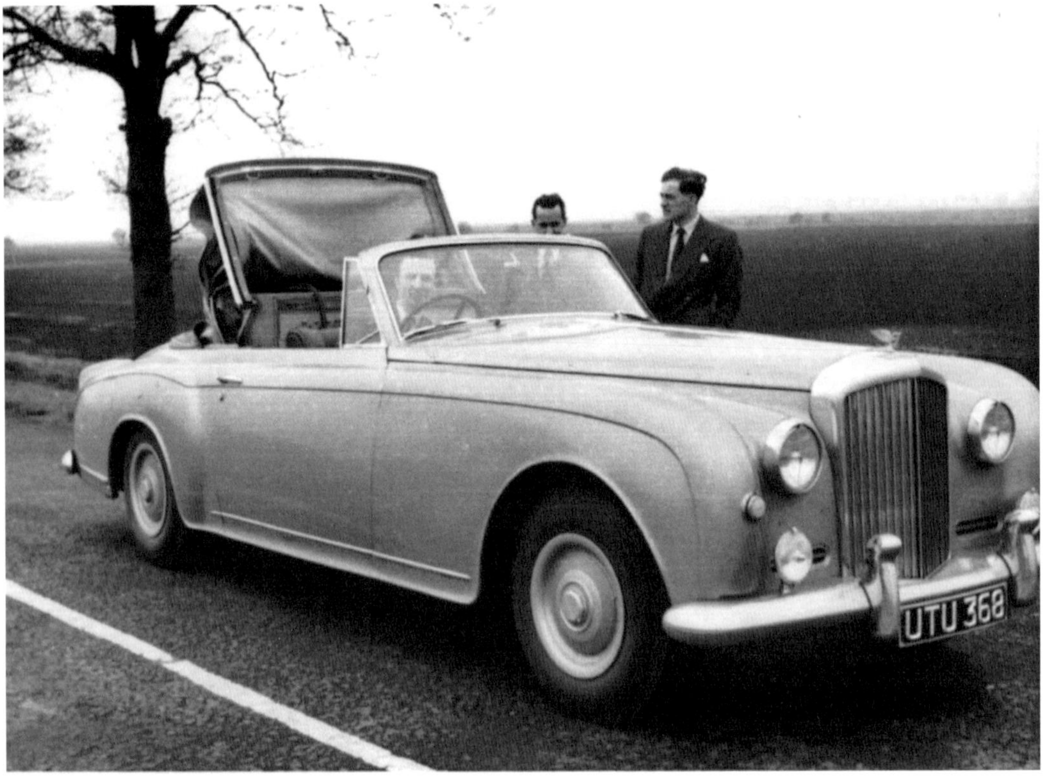

A Bentley S series convertible at Haxey, (Lincolnshire) in 1955. Standing are John Coyle and Derek Coulson, development engineers.

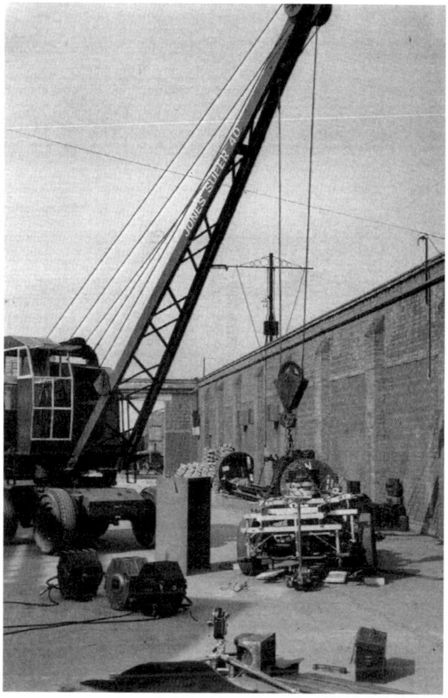

Right: Preparing to bump test an S series chassis behind the test beds. This crane was also used in an attempt to tow a mislaid Rolls-Royce off the embankment of the Chester railway line in 1965.

Below: Outside of the apprentices' school in 1964. A sectioned FB60 engine is being loaded onto the firm's Commer watched over by three engineering apprentices. The one on the truck is Brian Gabbatt, Raymond 'Taffy' Evans has his hands in his pockets.

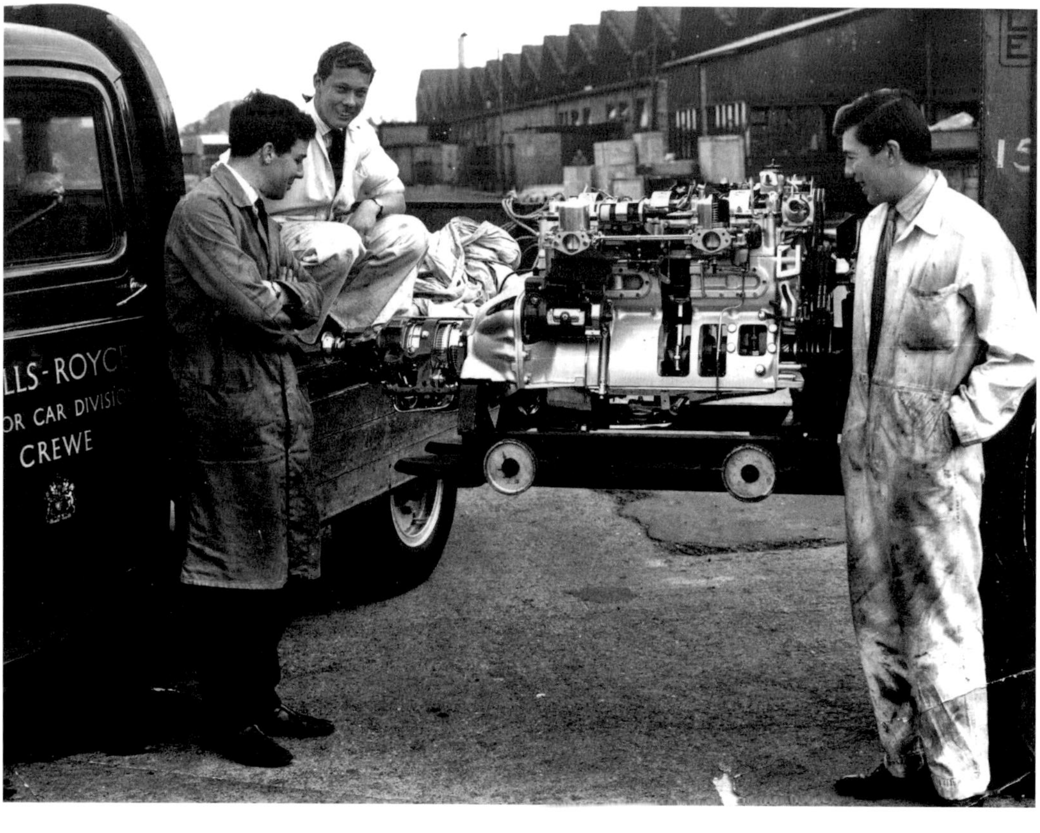

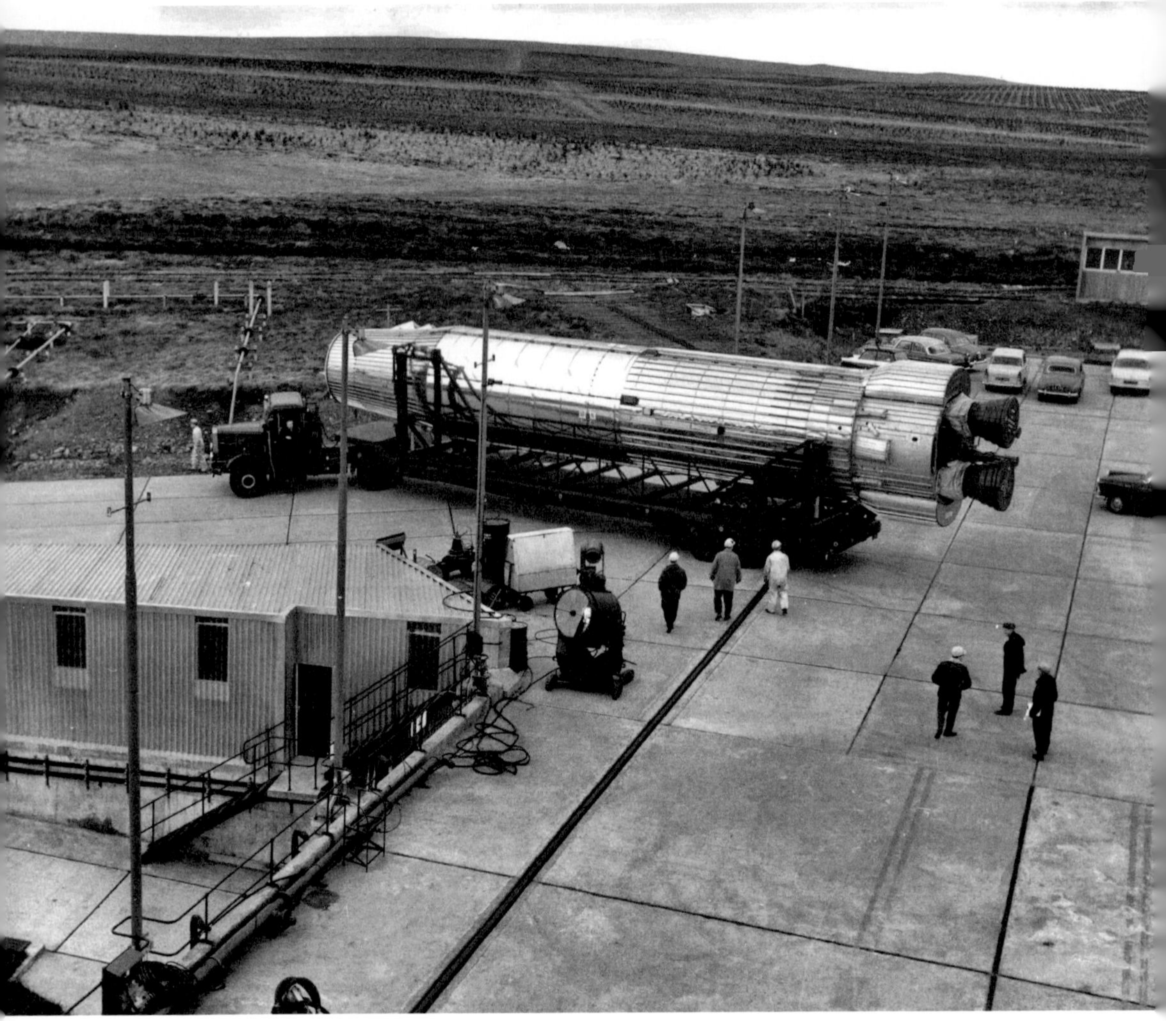

Pym's Lane also participated in the RR contribution to the Blue Streak rocket project. This image shows a rocket at the Spadeadam site in Northumberland.

five

FLOATING ON THE SHADOW

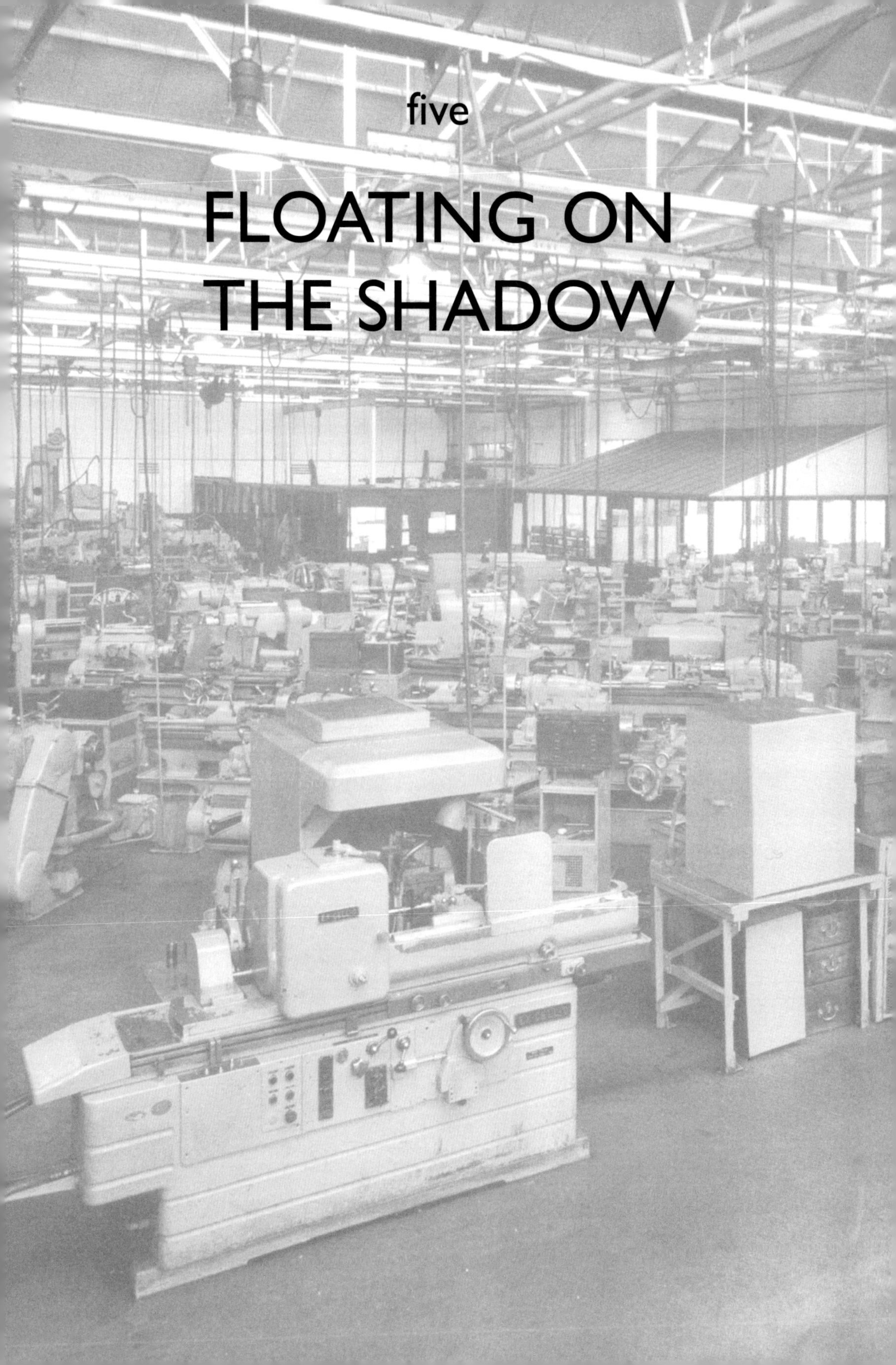

The Rolls-Royce Silver Shadow and Bentley T series, first offered for sale in 1965, were major departures from all previous RR cars as they were designed without a chassis. The cars also incorporated many other changes which caused Llewellyn Smith to state that he was glad he would never have to go through such a process again. It was fortunate that the Silver Shadow was a good seller because in 1971 RR called in the Receiver which meant that RR Motors became an independent company when it was separated from the aero division at Derby.

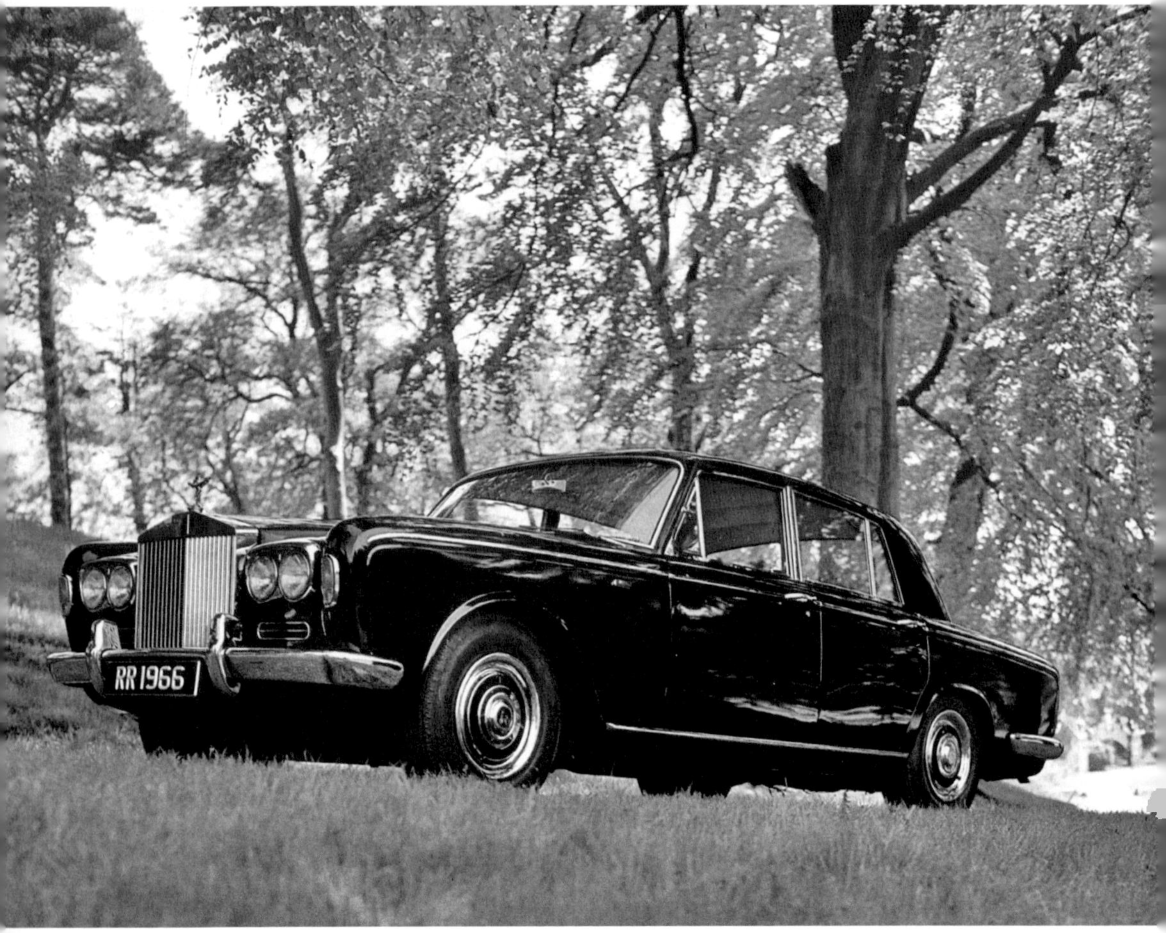

Rolls-Royce Silver Shadow in 1966 powered by the V8 engine first introduced in the Silver Cloud II series. Launch price £6,556. Eleven years later its price had risen to £19, 661.

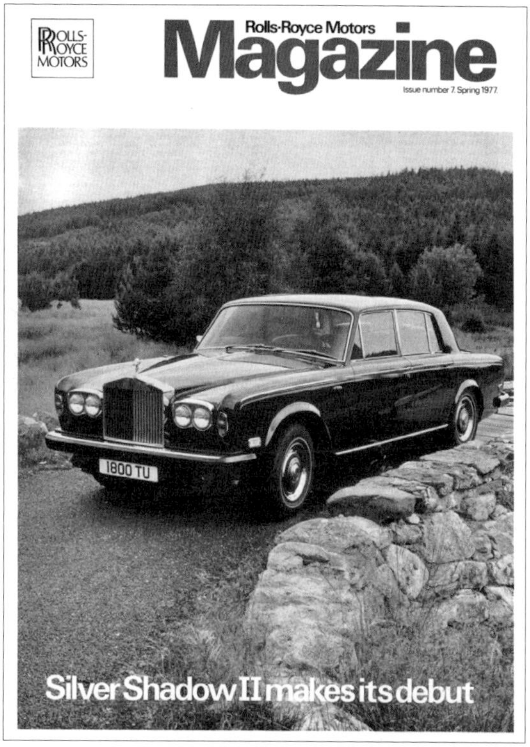

Right: Front page of the 'in-house' magazine announcing the arrival of the RR Silver Shadow II in 1977 priced at £22,810.

Below: Nearest to the camera is a Bentley Corniche first offered for sale in 1971 while next to it is a Bentley T series.

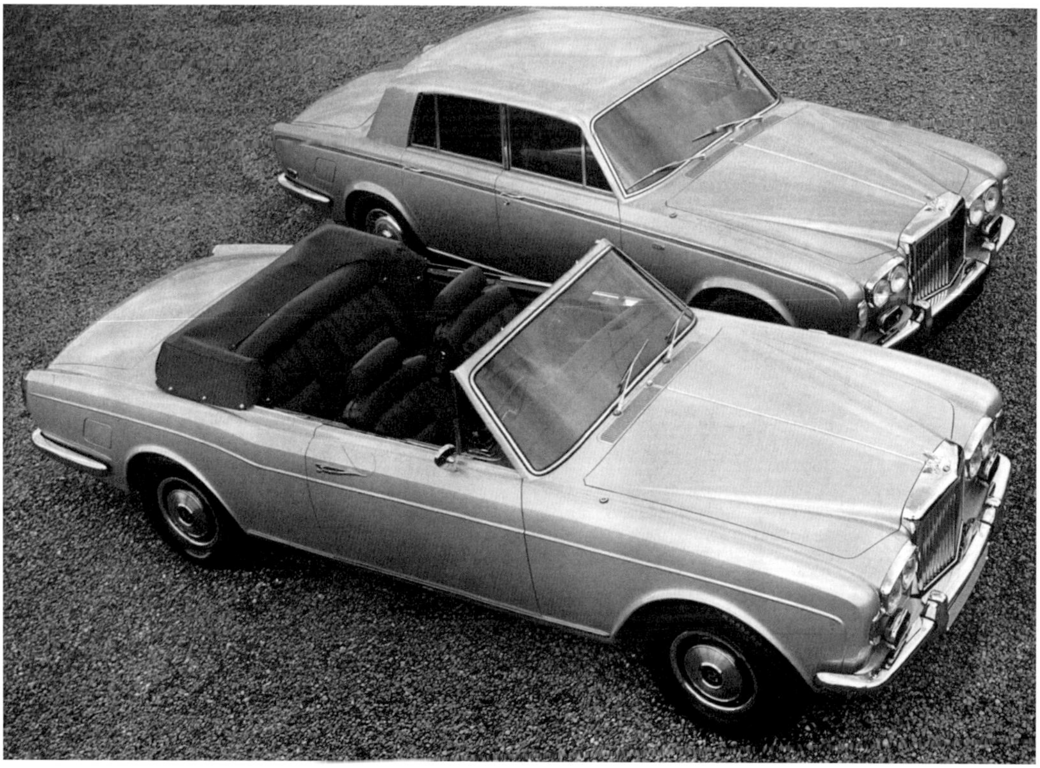

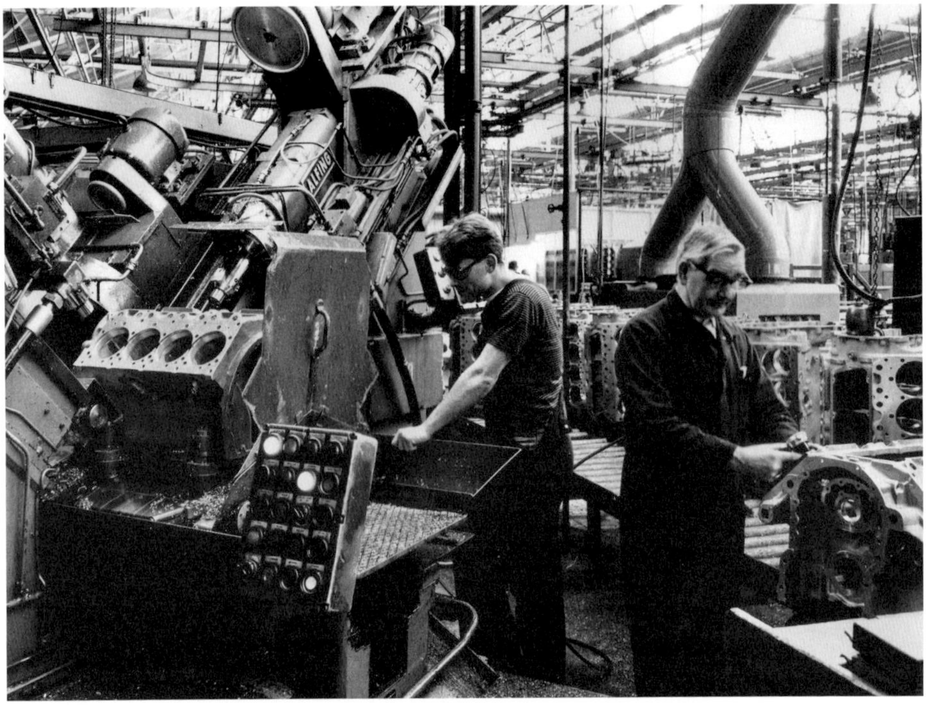

Drilling and tapping the crankcase for the cylinder head studs, 142 department (*c.* 1976).

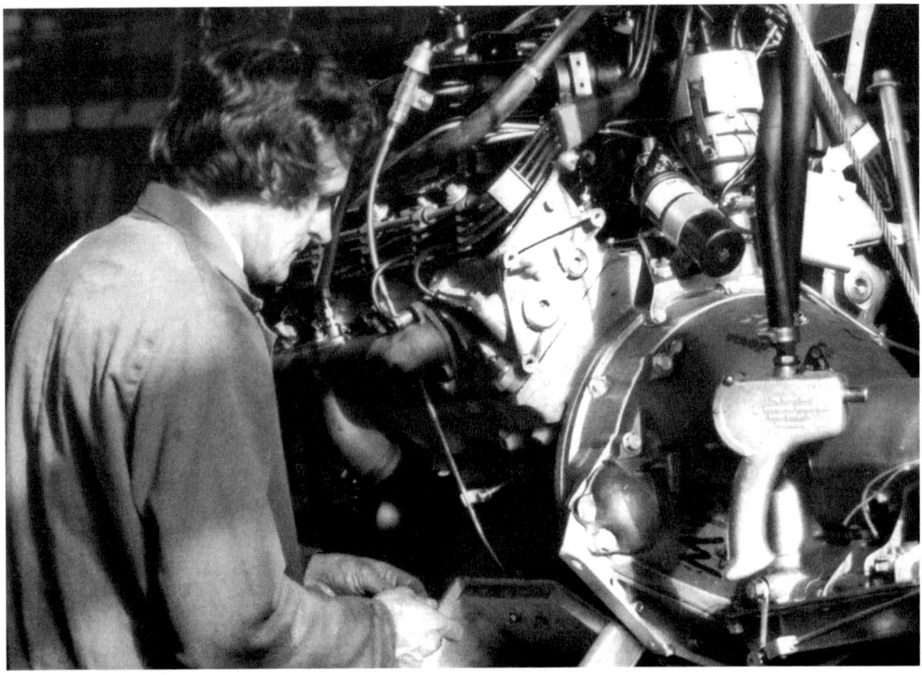

Gordon (Sam) Young fitting the GM 400 gear box onto a V8 engine. In 1953 he borrowed half a crown from the author to take a girl to the cinema.

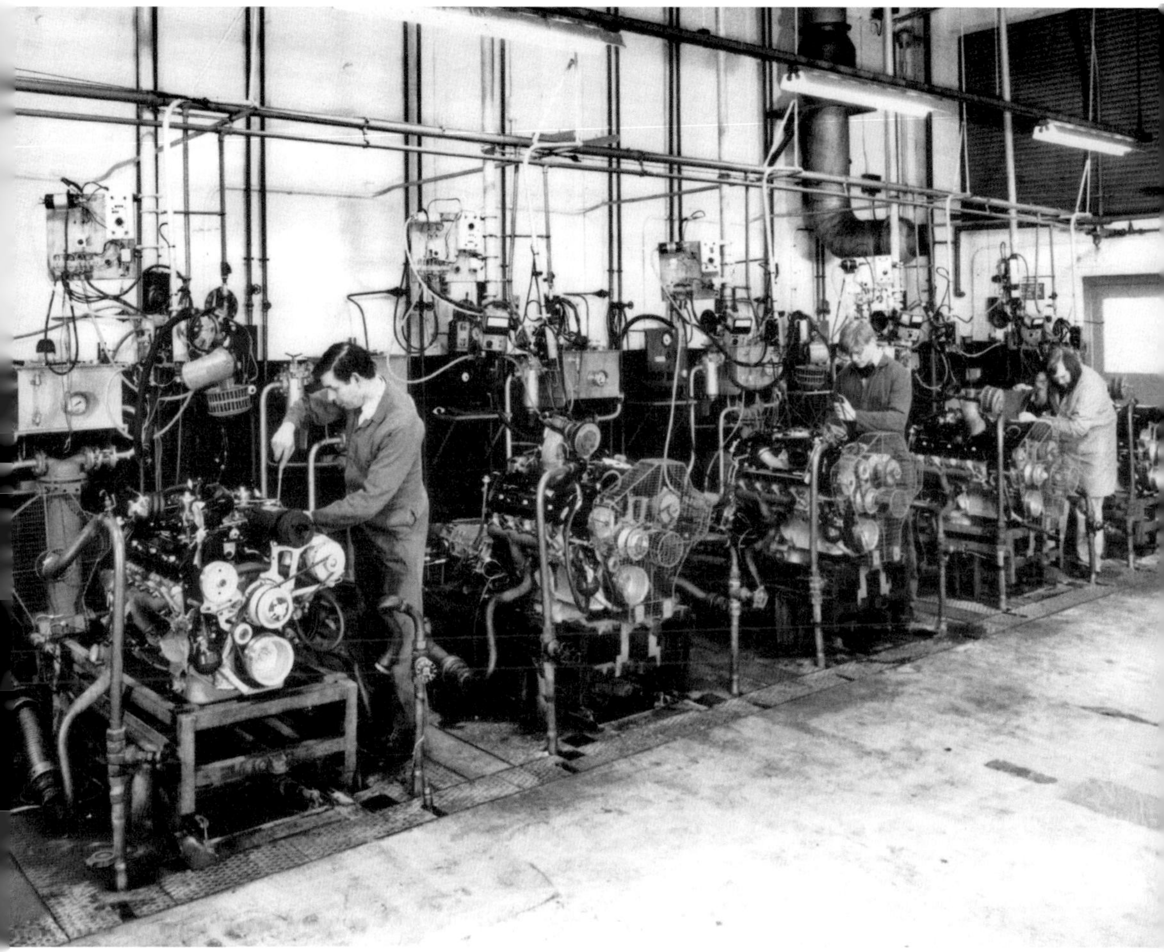

Most engines were run at 2,000 rpm for an hour powered by gas on the test beds. One in every hundred was tested at full power fuelled by petrol at a neighbouring test bed after which the engine was stripped down and every part inspected.

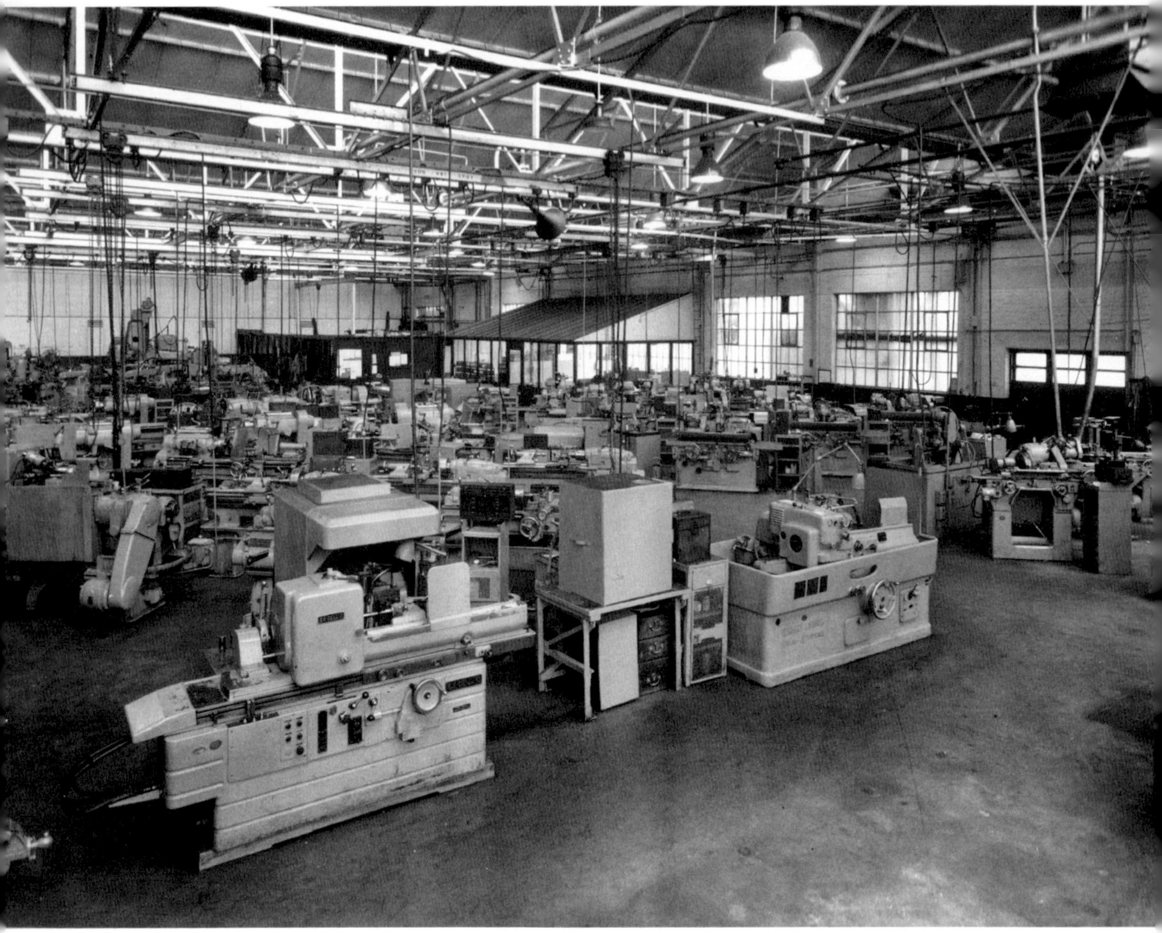

A rare view of the tool room looking south *c.* 1970 when the machines had just been painted. The partitioned area in the middle distance was the foreman's office. This department was not usually empty of employees as a night-shift and much overtime was worked.

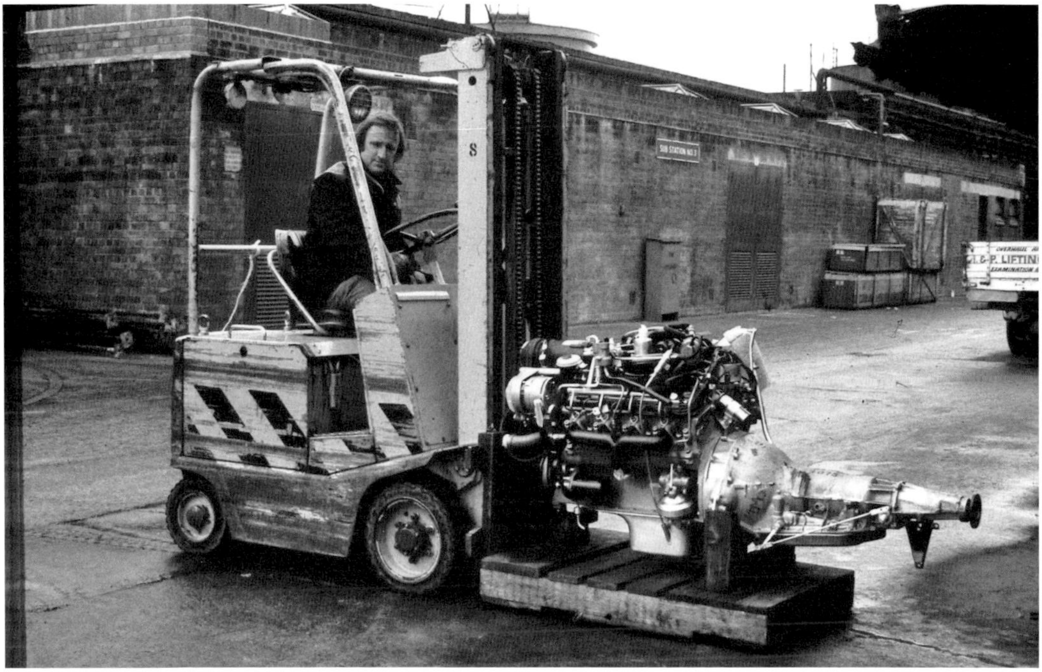

Silver Shadow engine, USA specification, being transported from the test bed area to the assembly line in the late seventies.

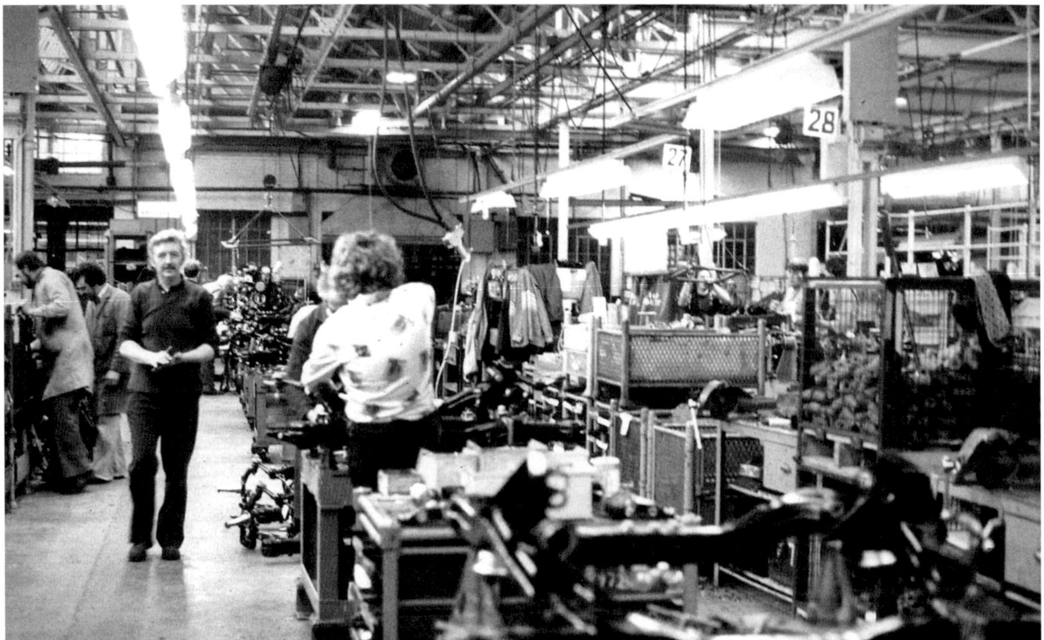

Front sub-frame assembly where the engine, gearbox, front brakes and steering linkage were fitted. Alan Ellis is walking towards the camera.

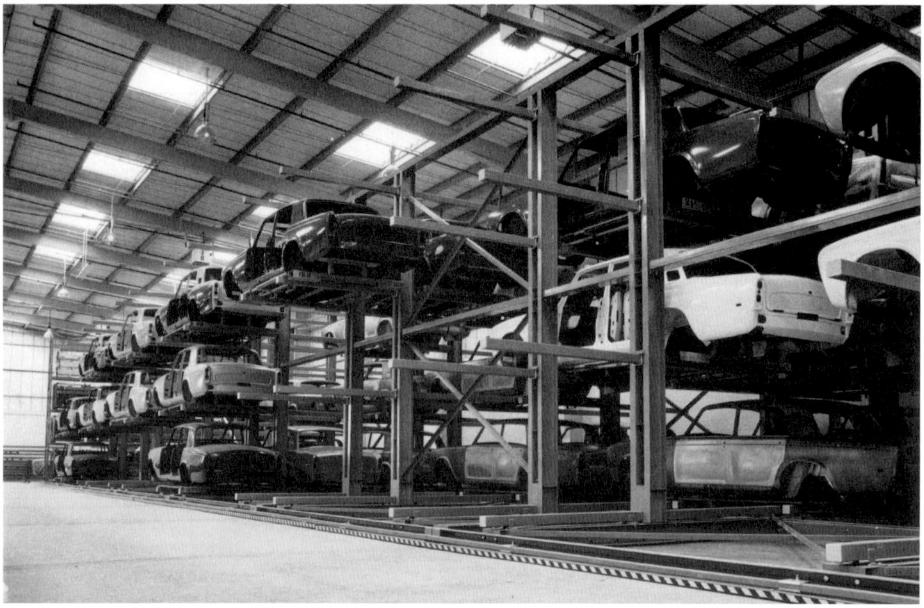

The body store for Silver Shadow II bodies. Notice that the bodies on the bottom rack are 'bodies in white' or pre-paint. The others were stored until needed on production.

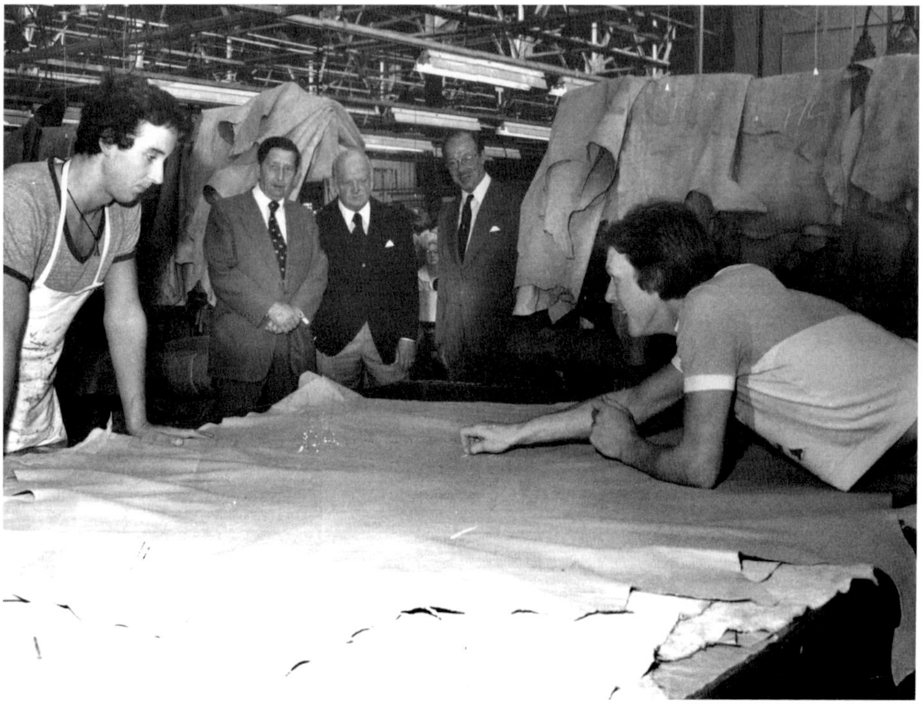

Cutting head-lining in the trim shop *c.* 1976. On the left of the group of onlookers is Tom Barlow, MD of the car division board who started at Pym's Lane in 1938 as a planner. Roy Jackson, production director is wearing spectacles.

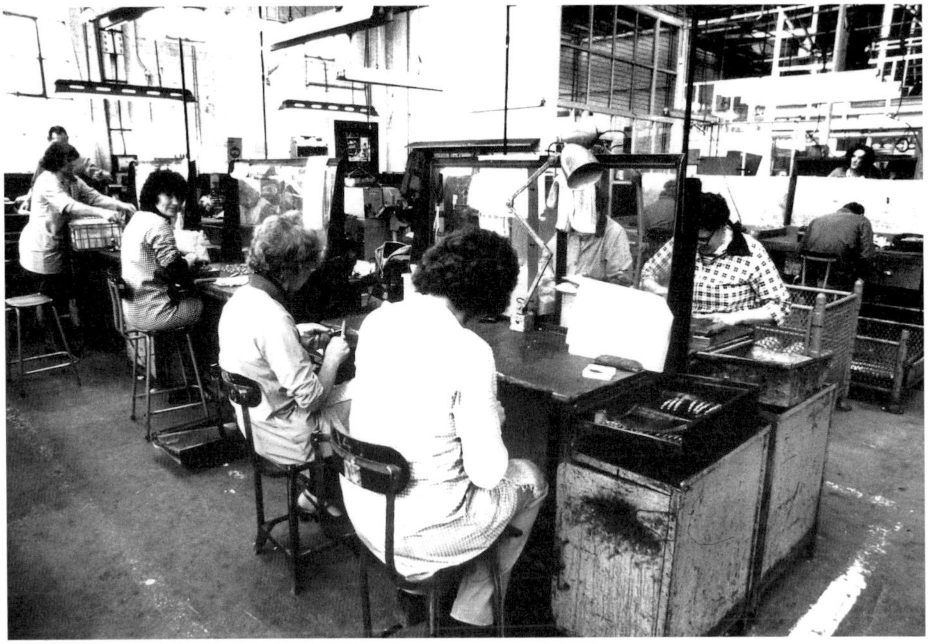

Women deburring (removing sharp edges) from small parts machined in the factory. This department was situated on the east wall of the main shop.

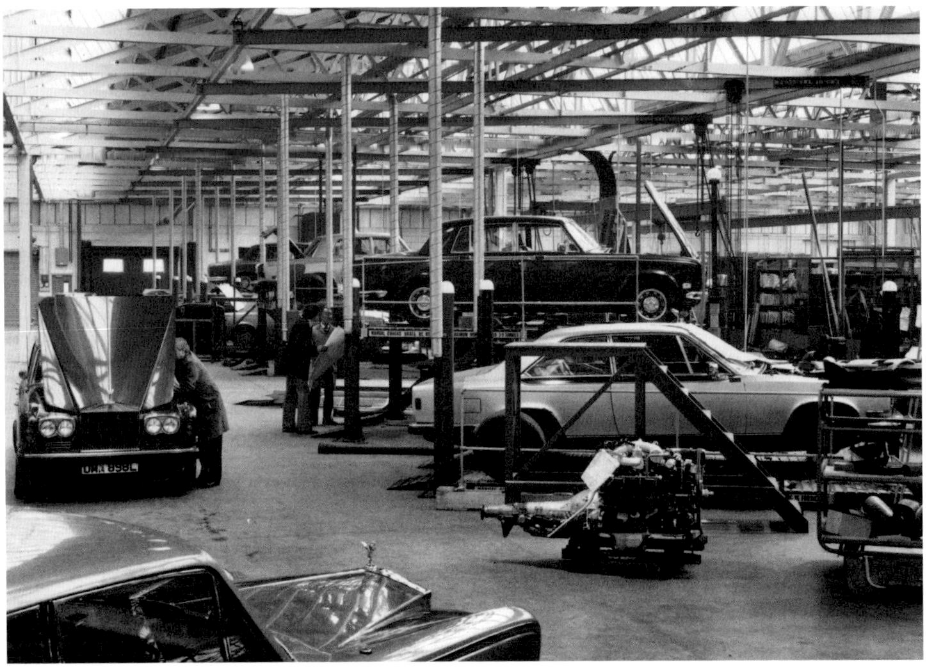

A general view of the experimental department's garage. The car in the centre nearest to the camera is the first Camargue ever made. In 1975 its purchase price was £29,250. Altogether around 530 were sold.

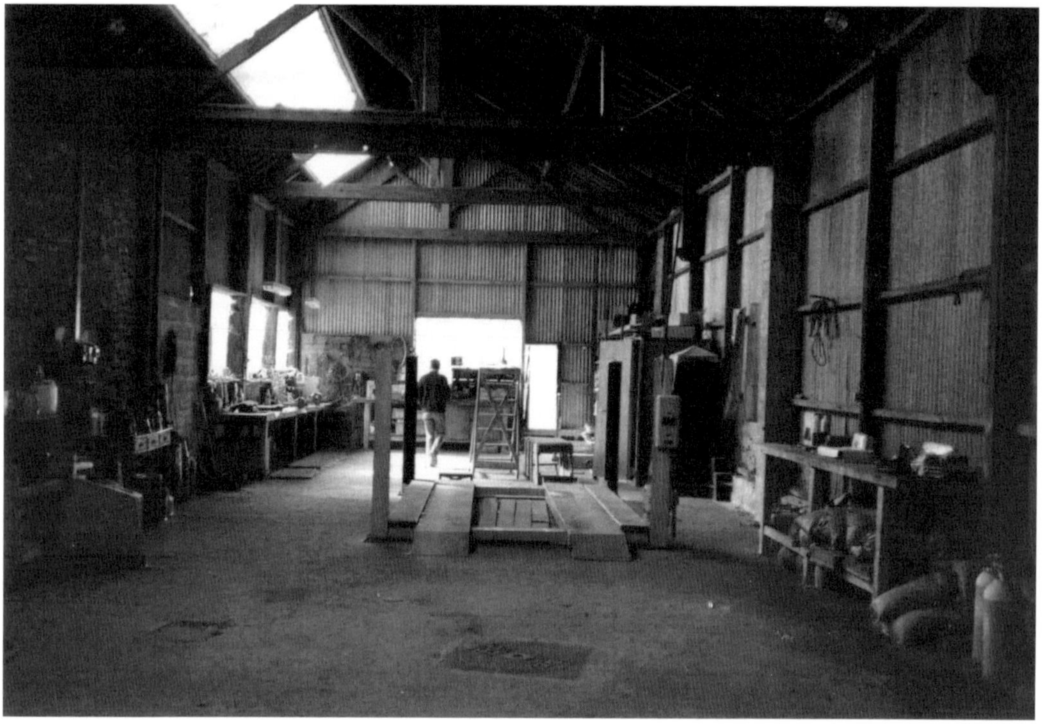

As much testing of prototypes was carried out in France the experimental department rented this small workshop at Blois.

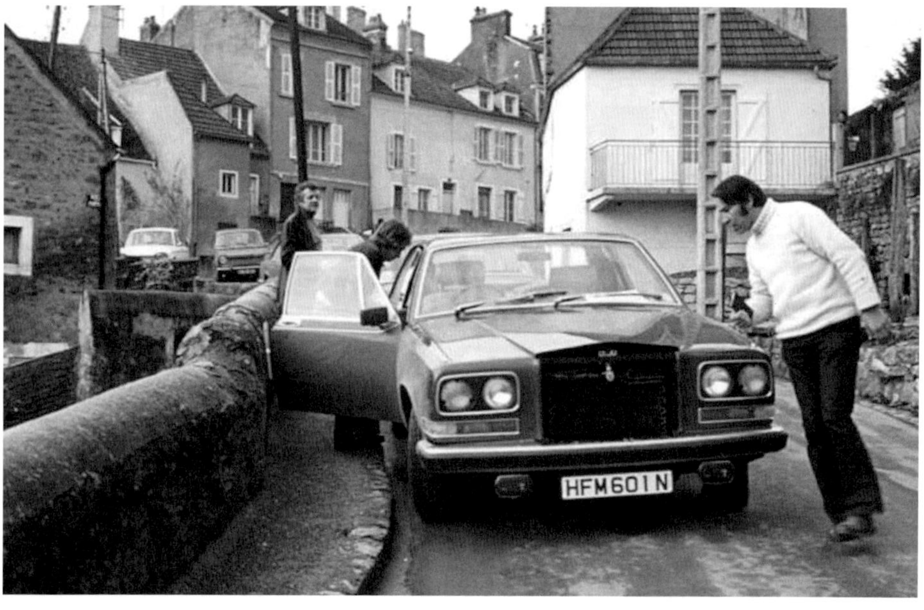

9 January 1975 between Le Havre and Avallon with a RR Camargue. Walter Lea, (test driver) is by the wall while Nick Colbourne (development engineer) is nearest the camera.

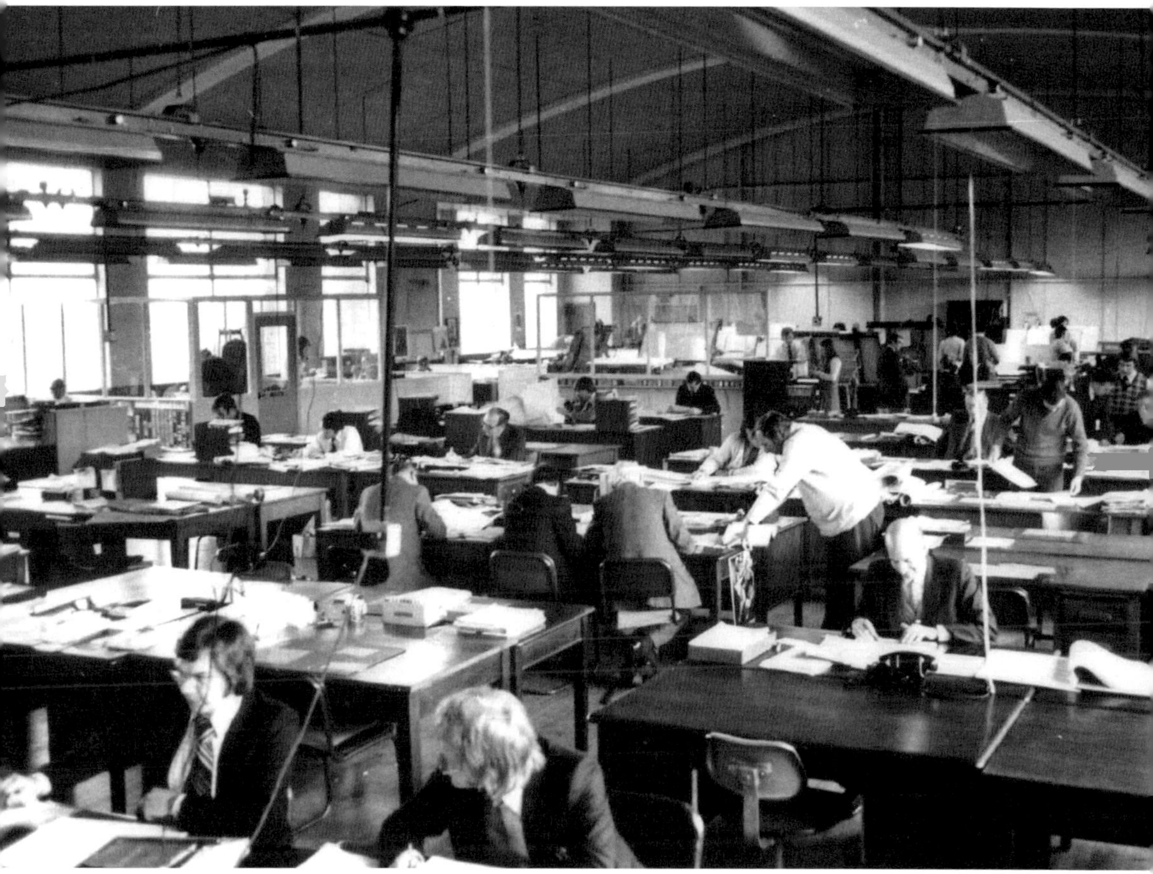

The planning office in the early seventies. The drawing boards in the background belong to the jig and tool draughtsmen.

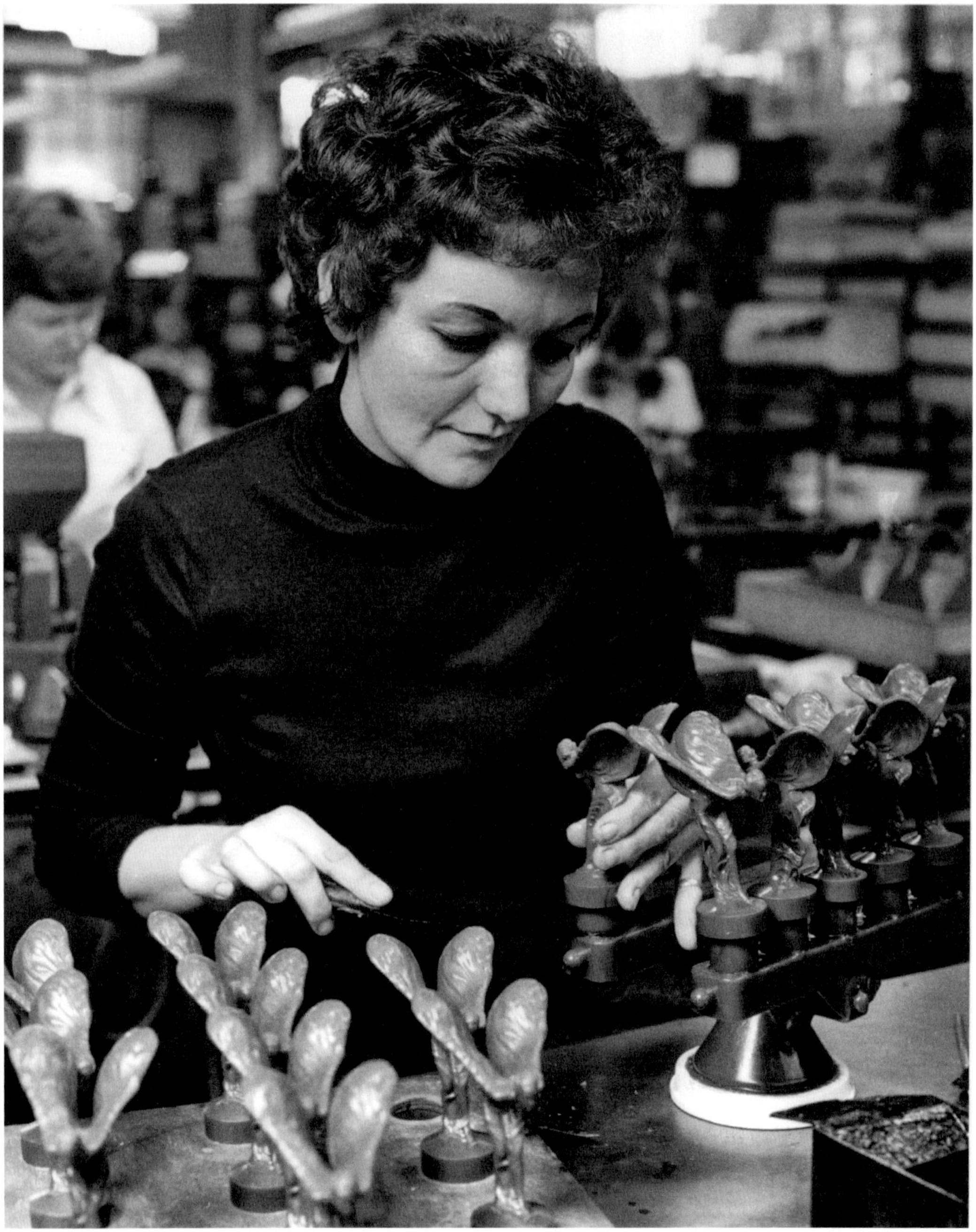

The investment foundry made items such as hip replacement joints, beer barrel bushes, and nozzle guide vanes for jet engines. The flying lady mascot that graced the famous radiator was also manufactured in the foundry.

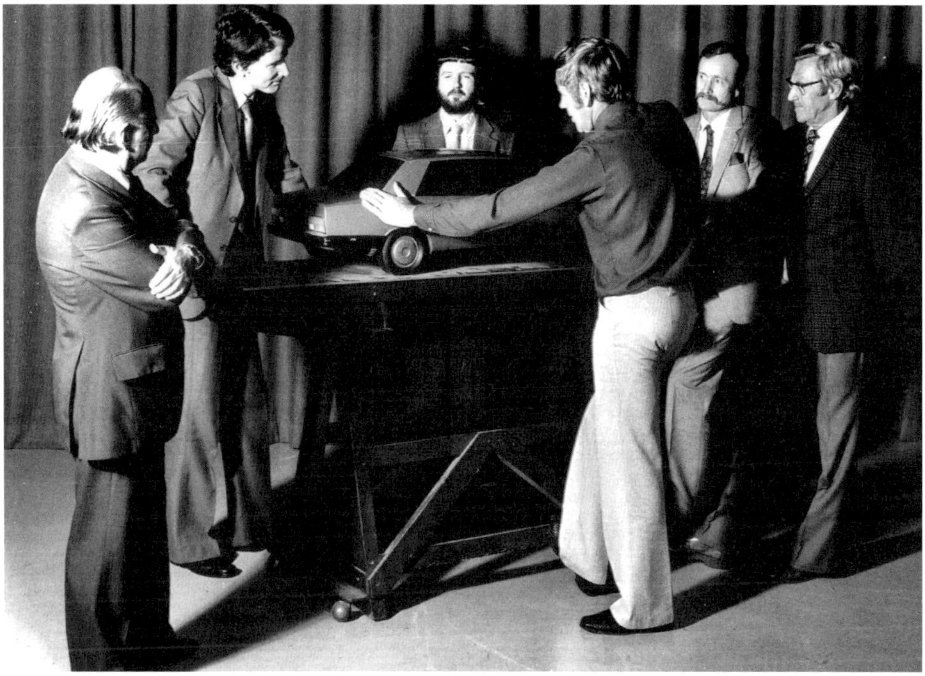

In the styling studio with a scale model of the Silver Spirit are Fritz Feller, Graham Hull, Ryan Lewis, Ron Maddocks, Martin Bourne and Norman Webster (*c.* 1974).

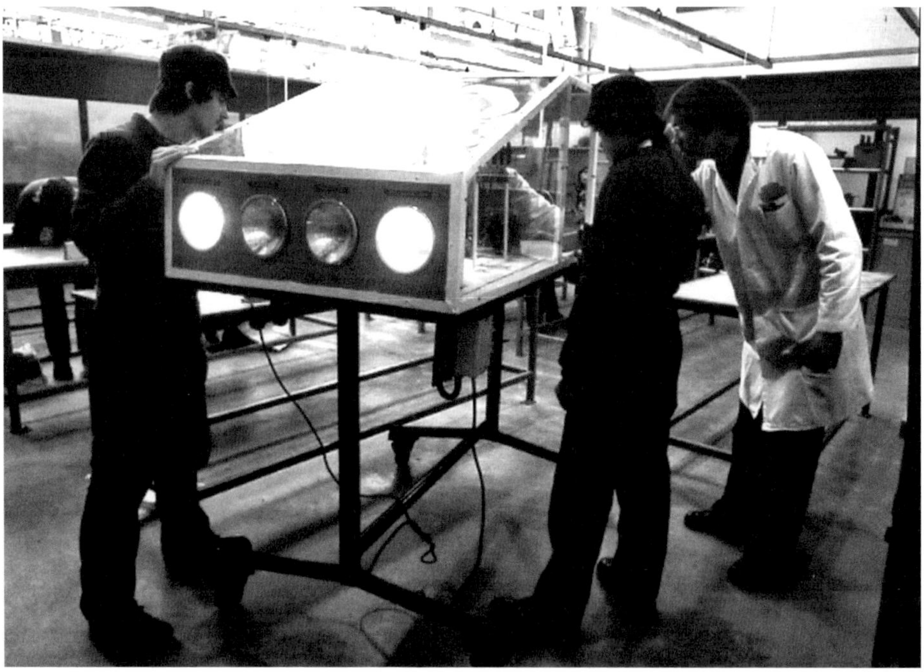

Apprentices' training school (*c.* 1980) where Dave Stevenson is passing on his knowledge of electronics.

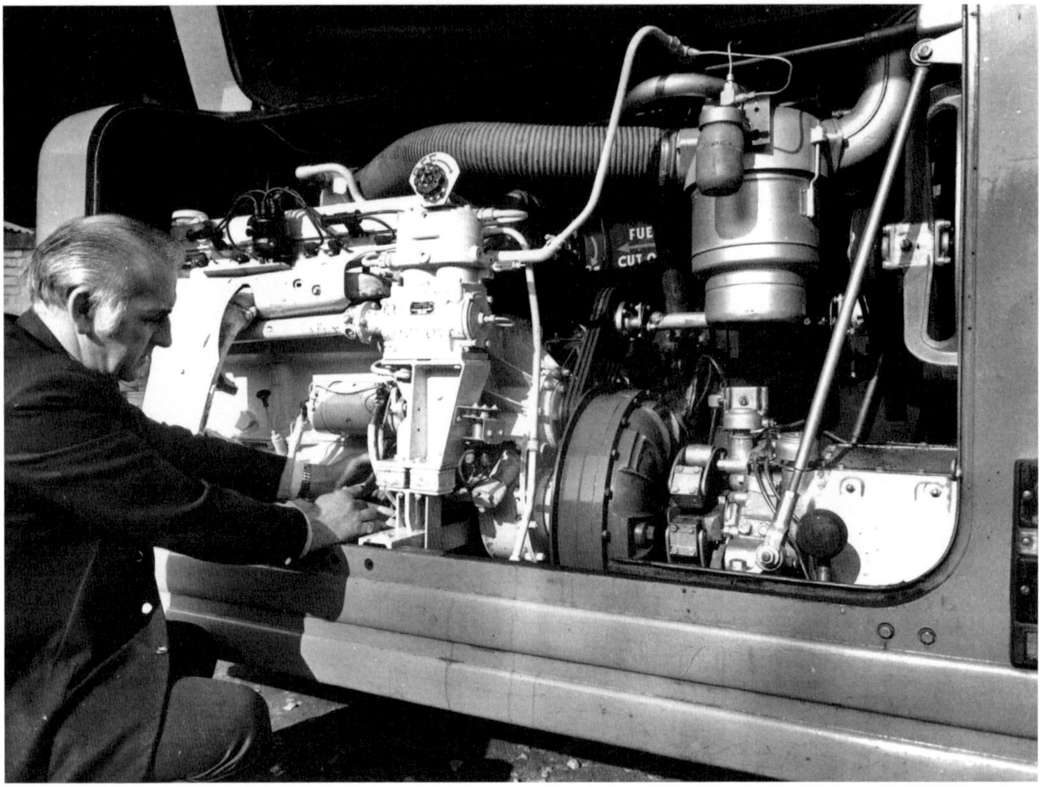

Not only motor cars were made at Pym's Lane. Here, in 1973, Jack Edwards, an experimental test driver, is examining a B81 engine powered by liquid propane gas.

six

MERGED WITH VICKERS

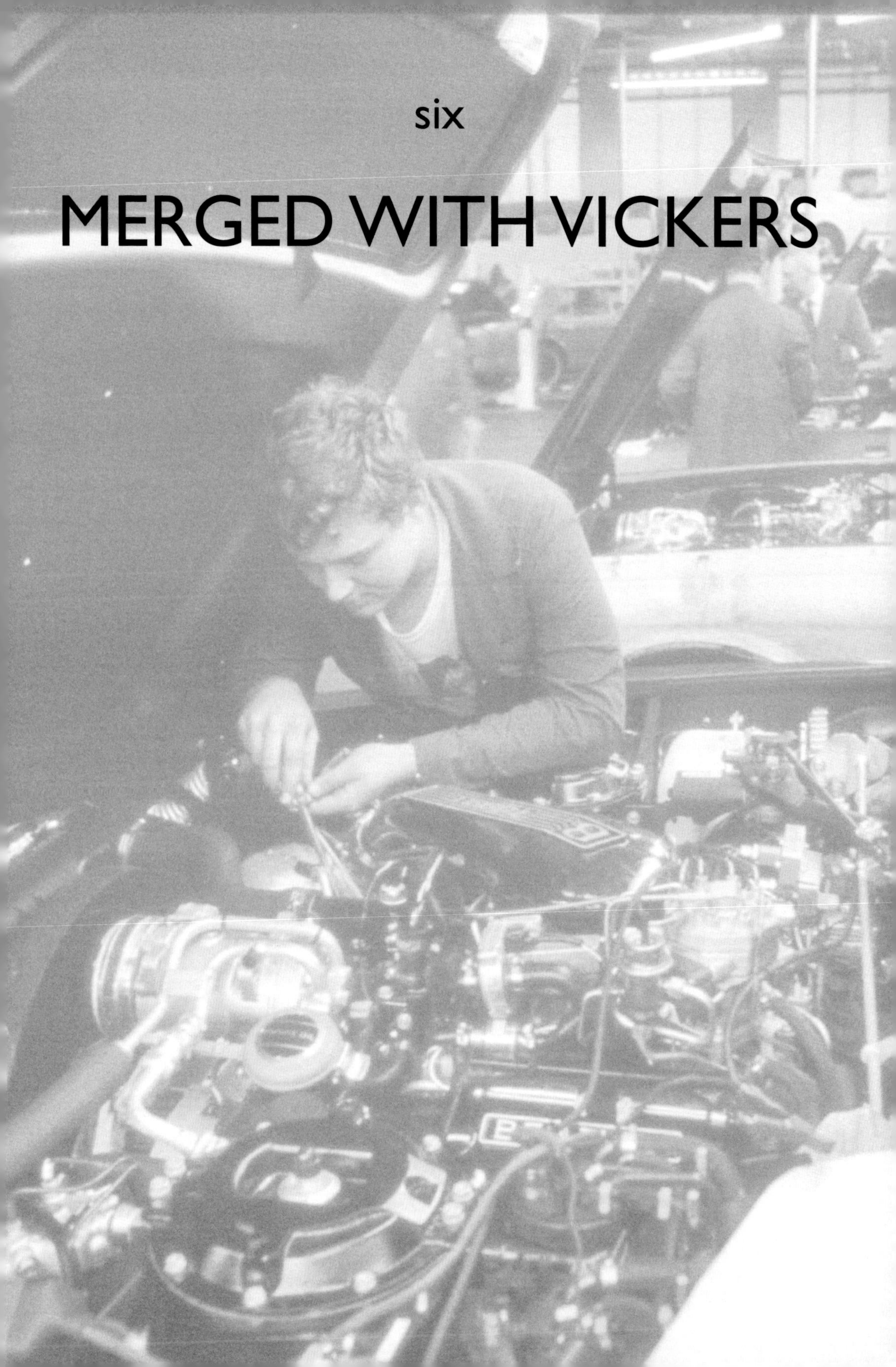

By the end of the seventies RRM was in need of a rich partner to help it through the financial maelstrom that was enveloping the British car industry. In June 1980 the announcement came that RRM was merging with Vickers where David Plastow was a non-executive director. In October of that year the SZ project matured in the shape of the RR Silver Spirit, a car that with the Silver Spur, Corniche, Bentley Mulsanne and Turbo R carried the firm until the close of the nineties.

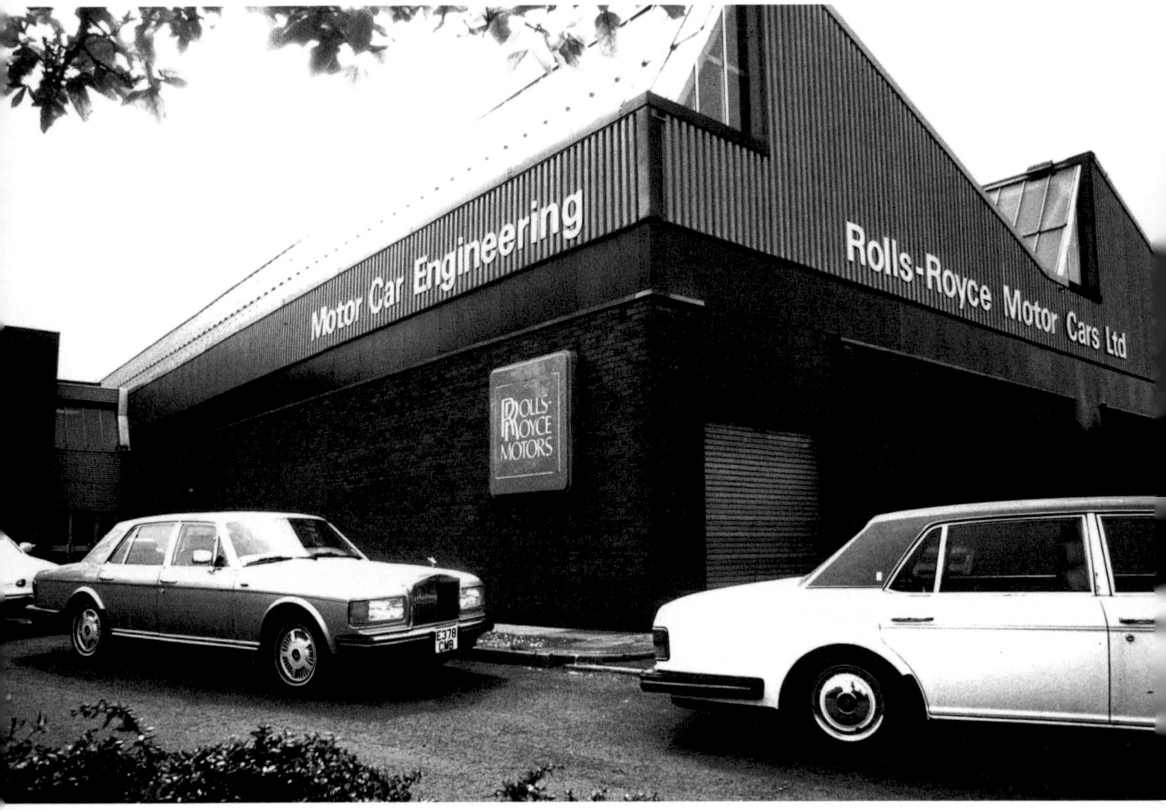

A Silver Spirit and Silver Spur outside of the engineering block (*c.* 1986). In 1987 a Silver Spirit cost £73,080 whereas a Silver Spur was £87,189.

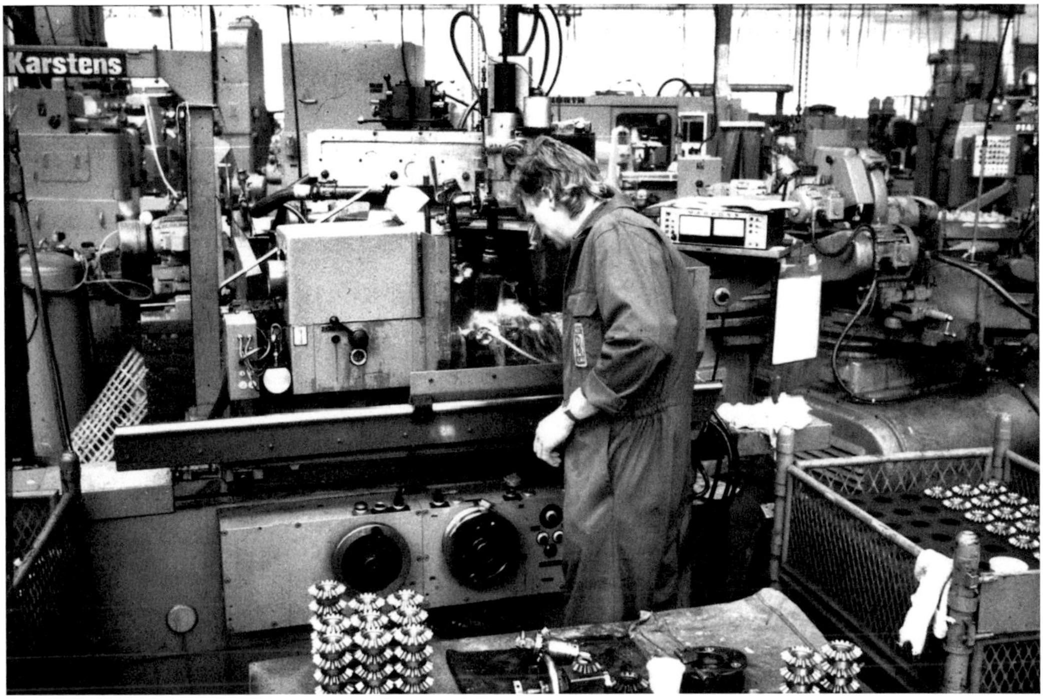

Above: Machining bevel gears for the differential train that fitted into the rear axle.

Right: Whenever capital was available new machines were purchased to speed up production. Here can be seen the Huller Hille 150 machining centres installed at great cost at the end of the eighties.

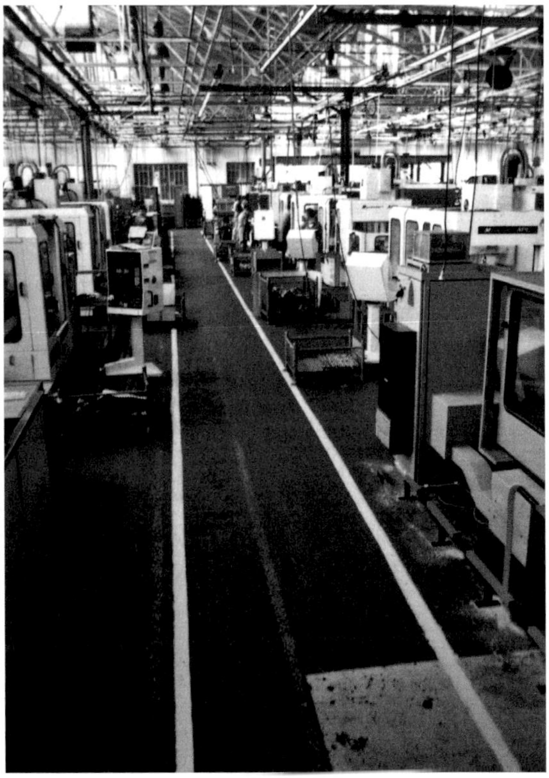

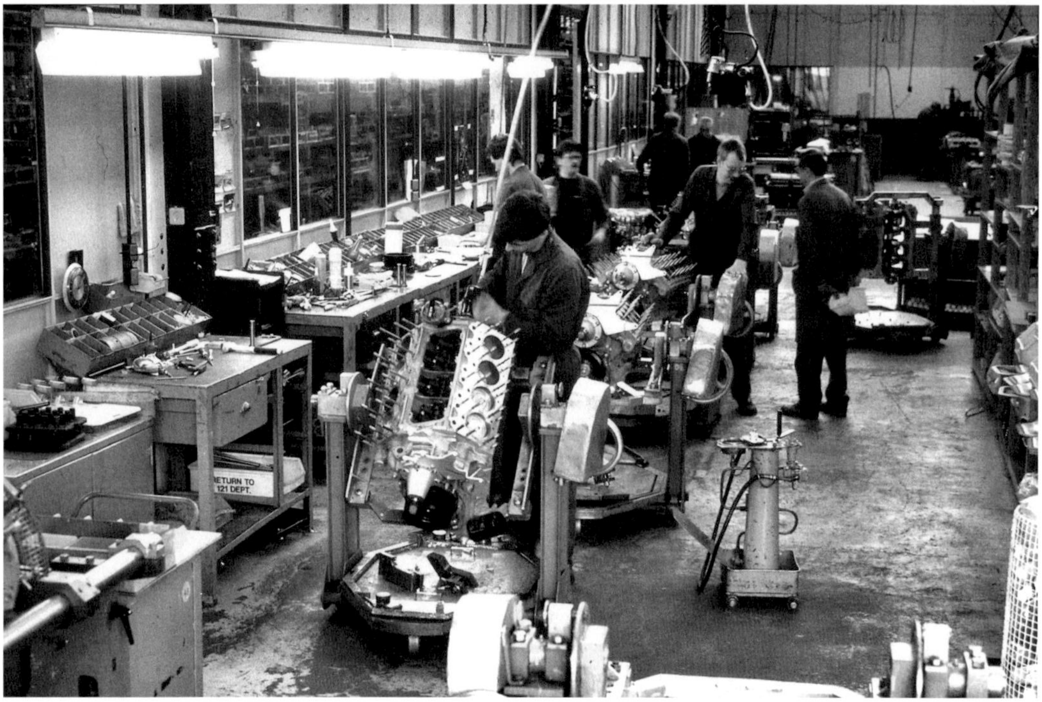

Part of engine build (*c.* 1988).

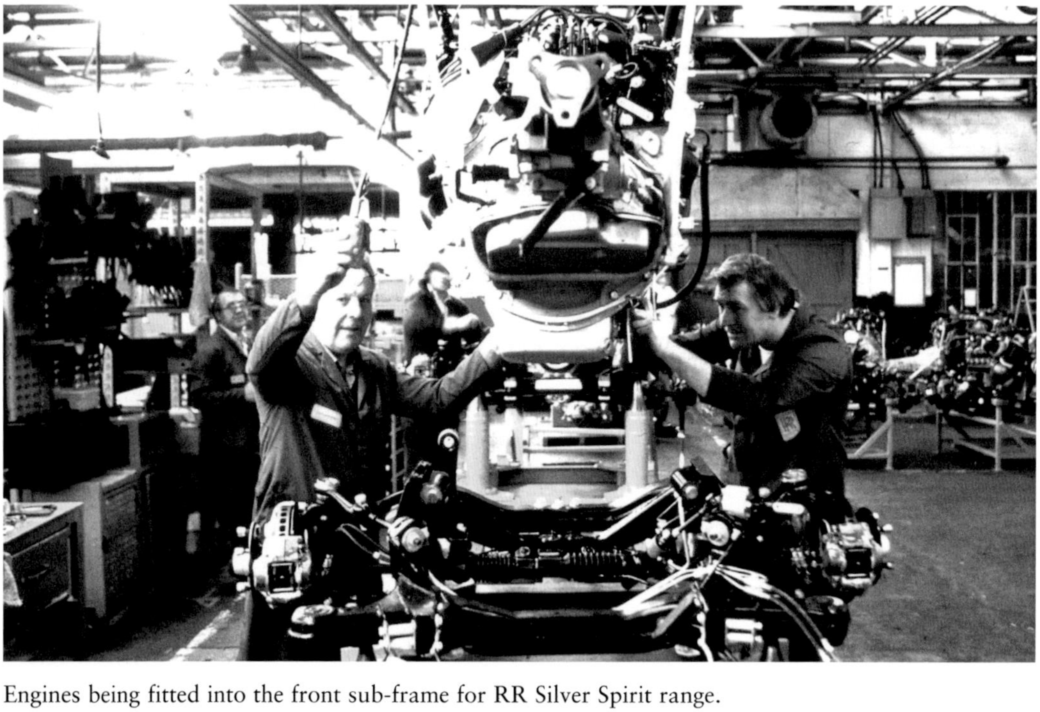
Engines being fitted into the front sub-frame for RR Silver Spirit range.

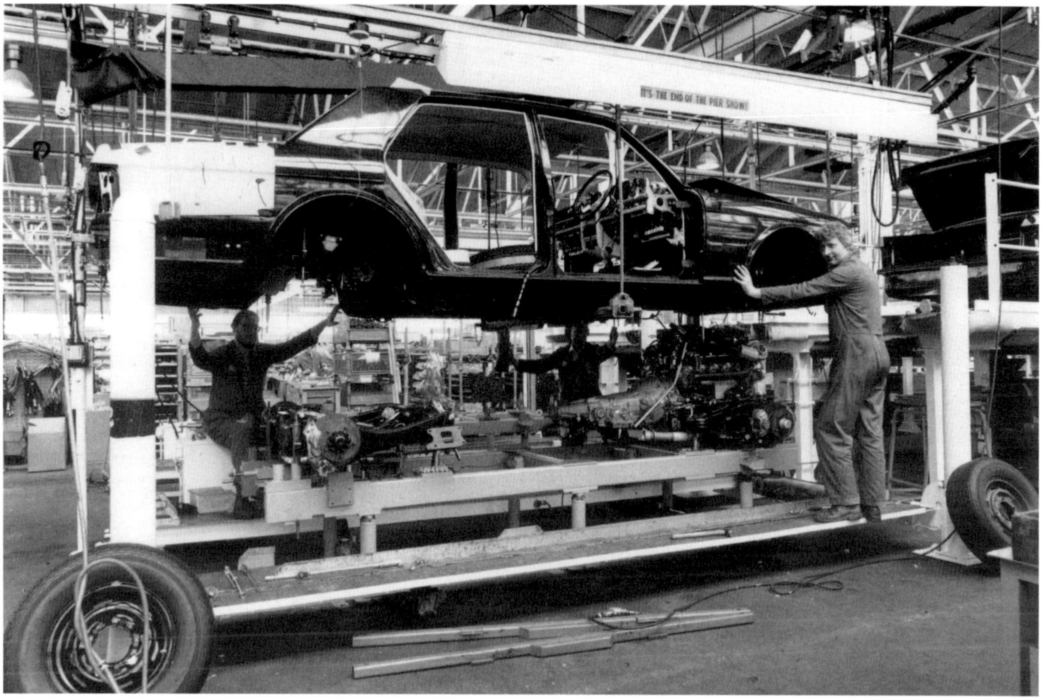

A Silver Spirit being mounted onto its sub-frames. Brian Stockton is at the front.

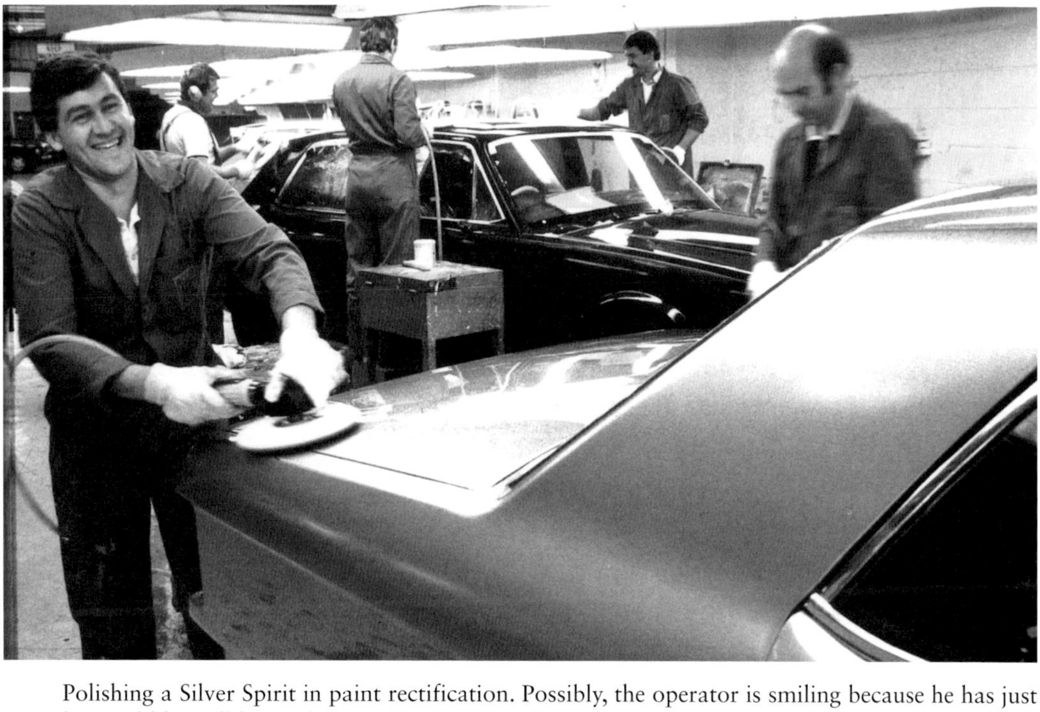

Polishing a Silver Spirit in paint rectification. Possibly, the operator is smiling because he has just been told he will be wotking on Sunday. Double time!

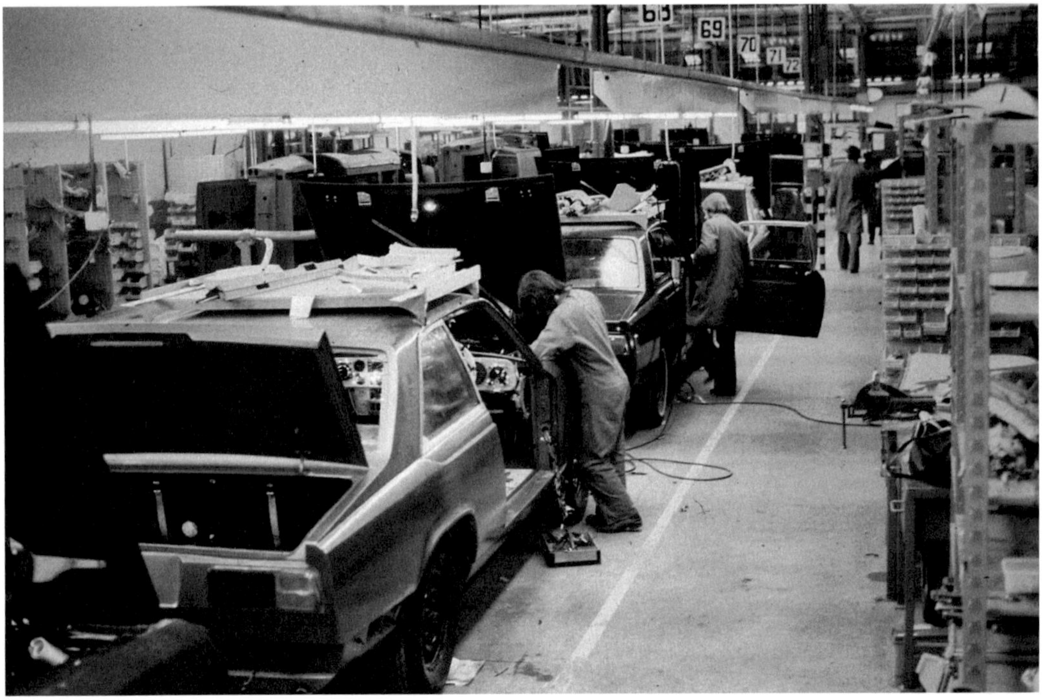

Camargue assembly line (*c. 1983*). Production of this model was concentrated at Crewe from 1978.

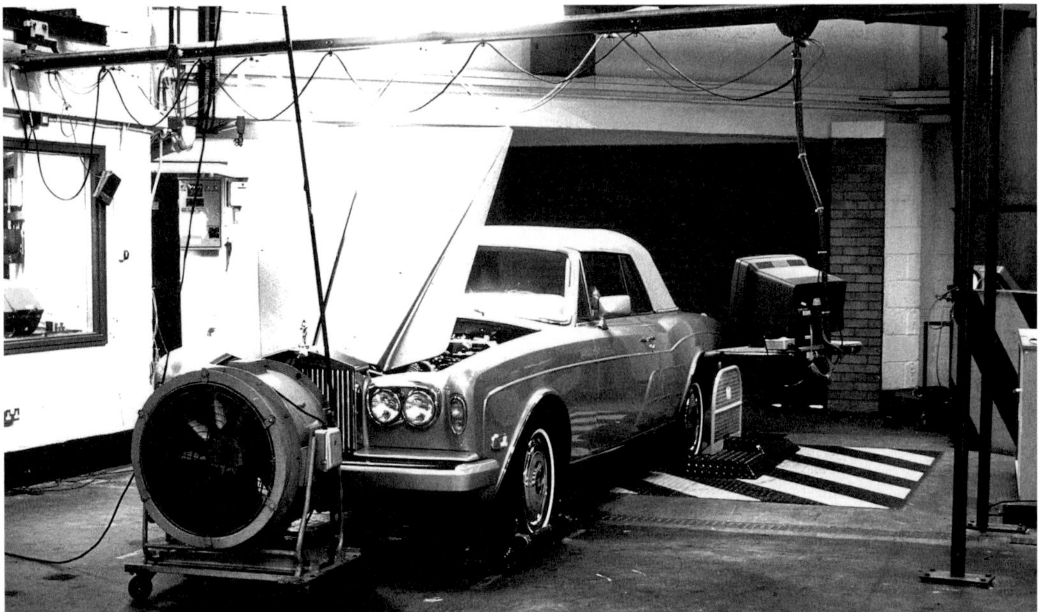

One of the great unsung achievements of the development staff was coping successfully with the flood of environmental and safety legislation. Here a RR Corniche (priced in 1989 at £110,543) is undergoing an exhaust emissions test.

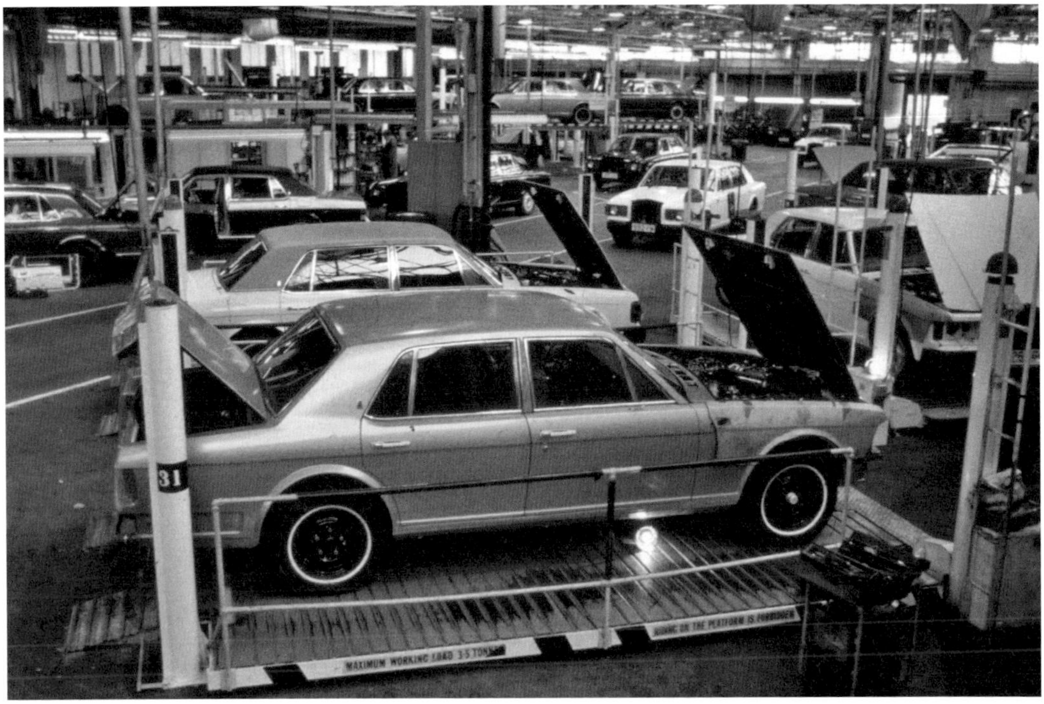

Road test general view. The nearest car is a Silver Spirit (*c.* 1988).

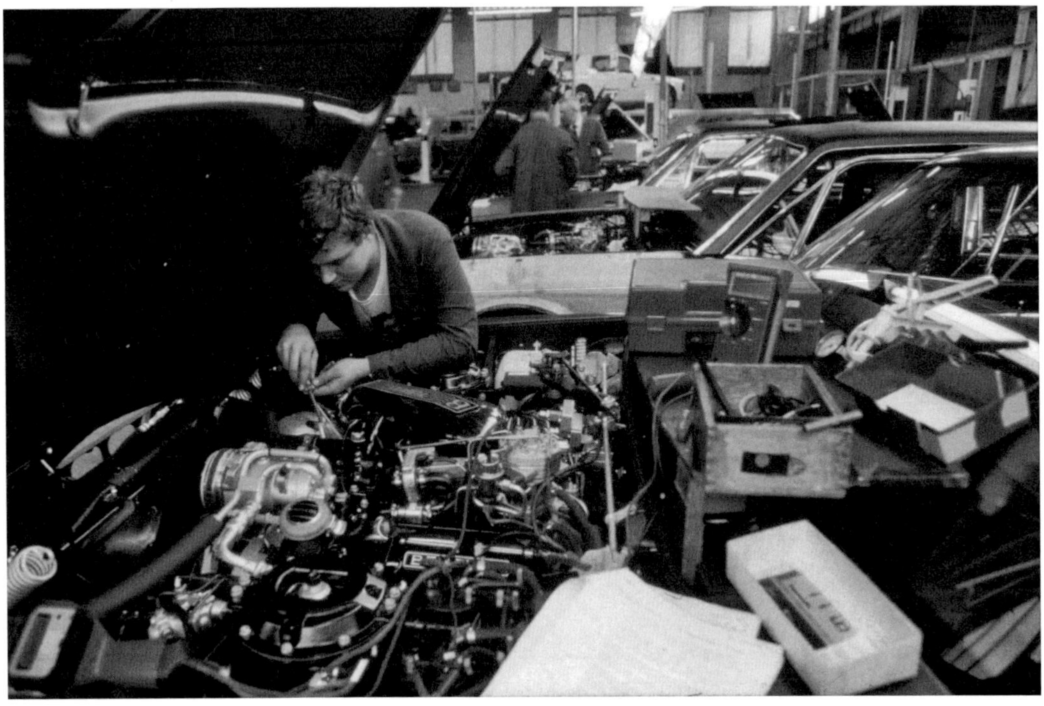

Rectifying minor faults revealed by road test on a Bentley Turbo R.

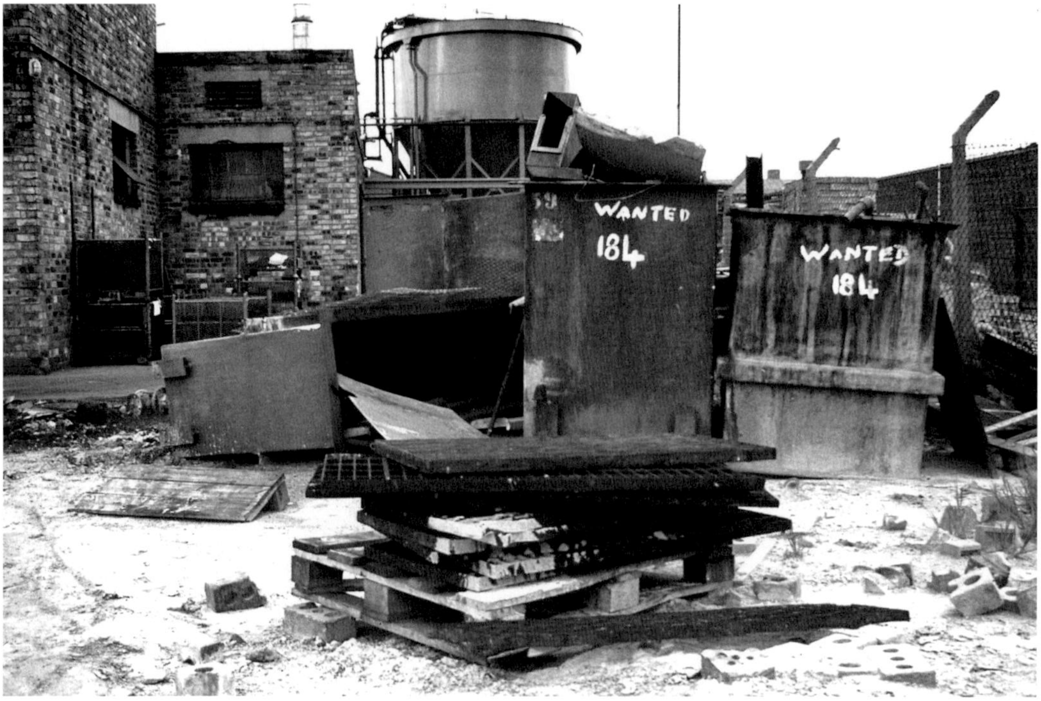

The rear of the oil stores in 1988 when a 'Think Tidy' campaign was organised.

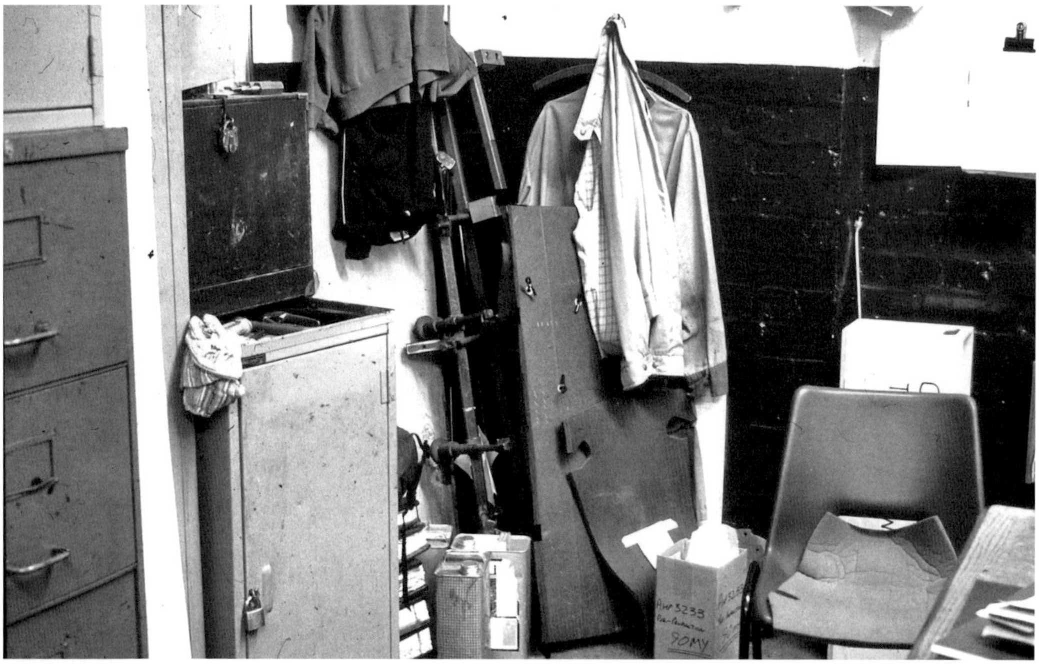

One of the workshop offices in need of tidying (*c.* 1988). Many important historical documents were destroyed at this time.

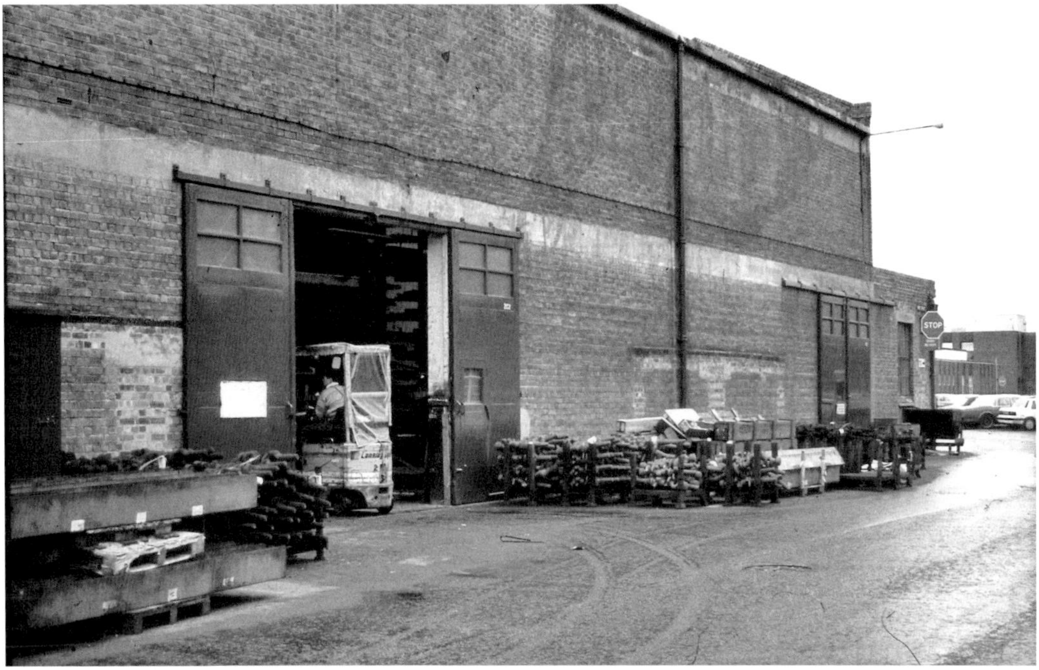

Steel stores (*c.* 1988). Notice the vestiges of wartime camouflage above the door.

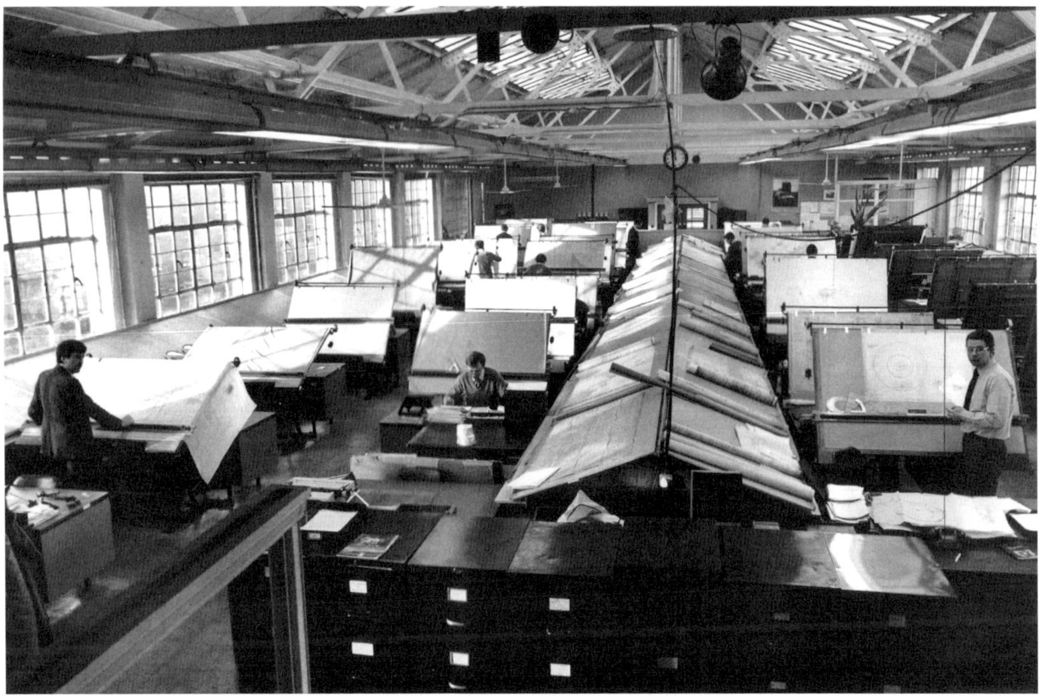

Design office (*c.* 1984). Nearest the camera are C. Burrows, M. Smith and M. Spice. Frank Tarleton's office was on the left, with George King's to the right.

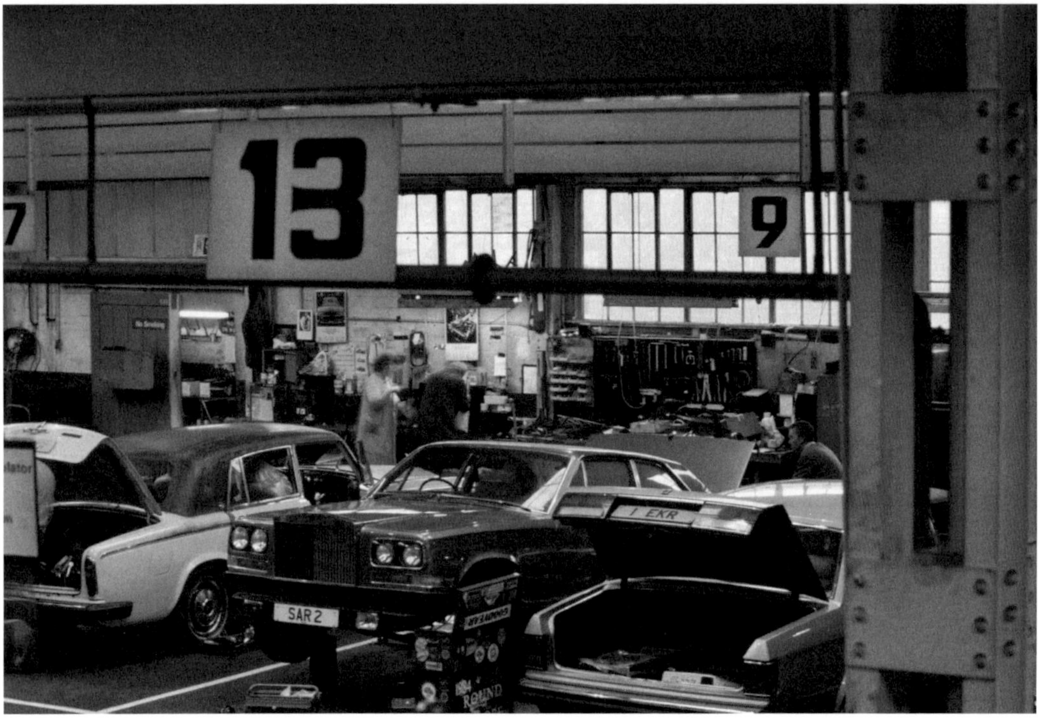

Car repair always had an interesting selection of cars (*c.* 1984). A multiplicity of trades and skills were needed in this department.

seven

THE END OF THE ROAD

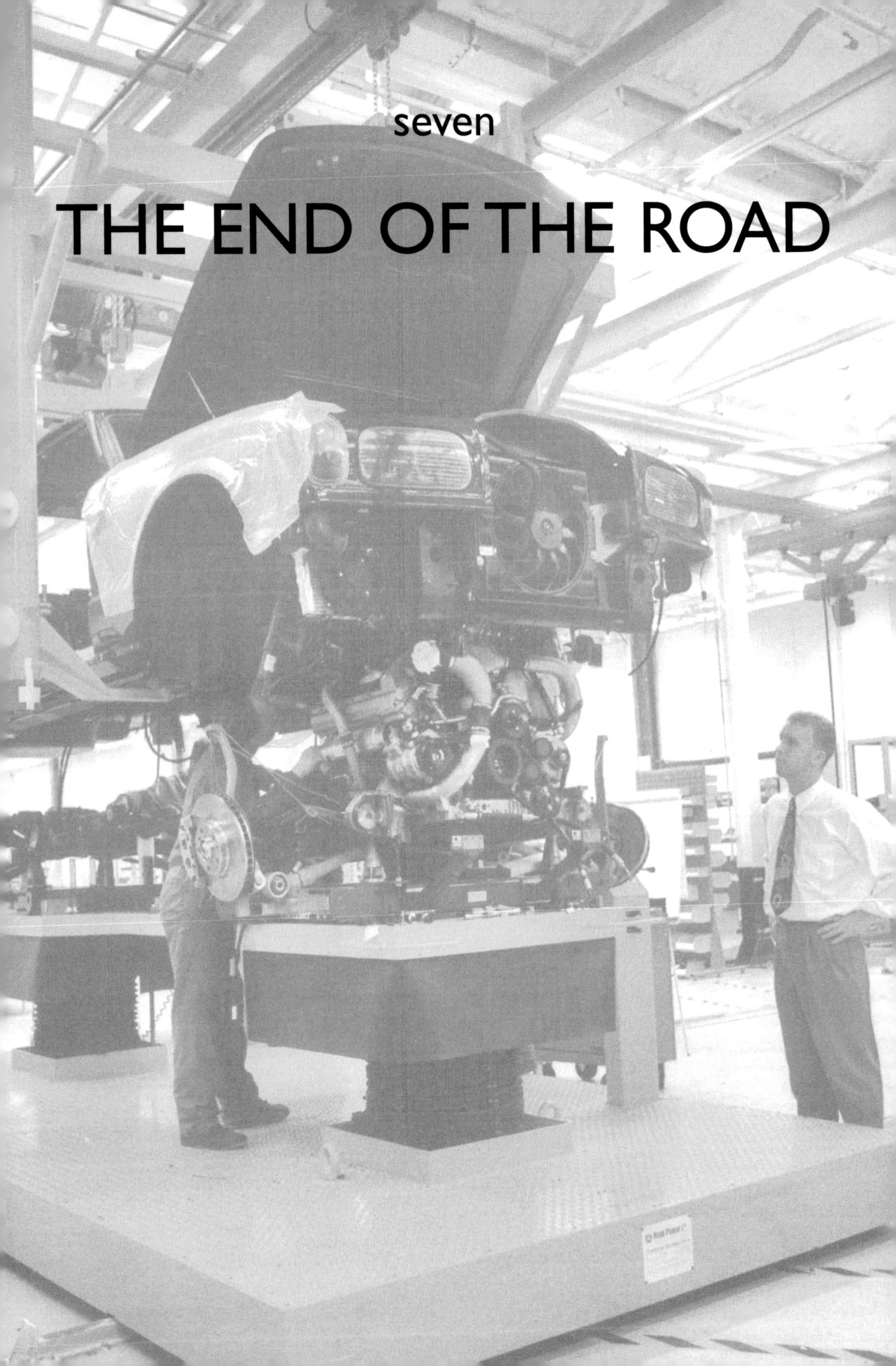

At the beginning of the nineties production of motor cars fell drastically from just over 3,300 cars to 1,723 with the consequential effect of compulsory redundancies. Despite this three new models were launched namely the Bentley Continental R, RR Silver Spur limousine and the Bentley Brooklands. Worse still in some ears was the news that the firm was up for sale.

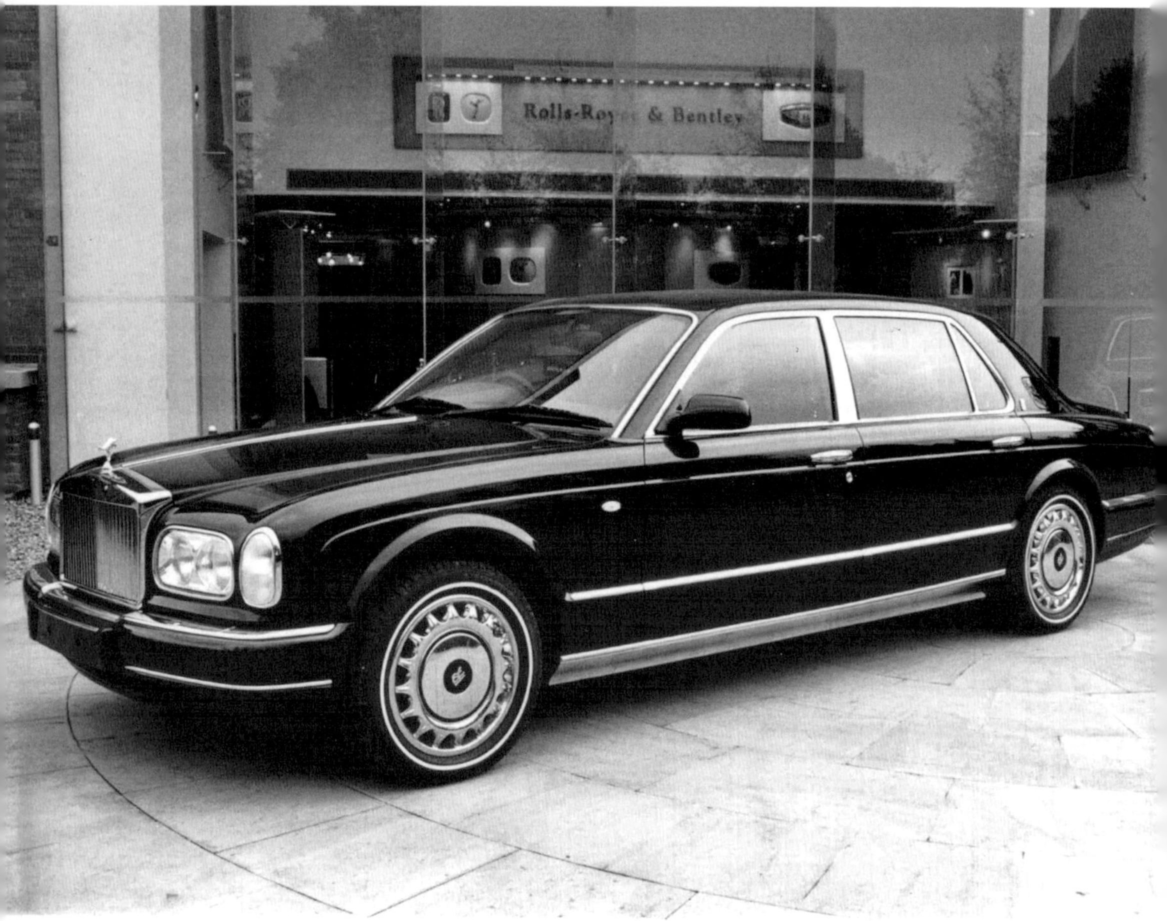

Rolls-Royce Silver Seraph, launch price £155,000, outside of the restyled main entrance (*c.* 1998).

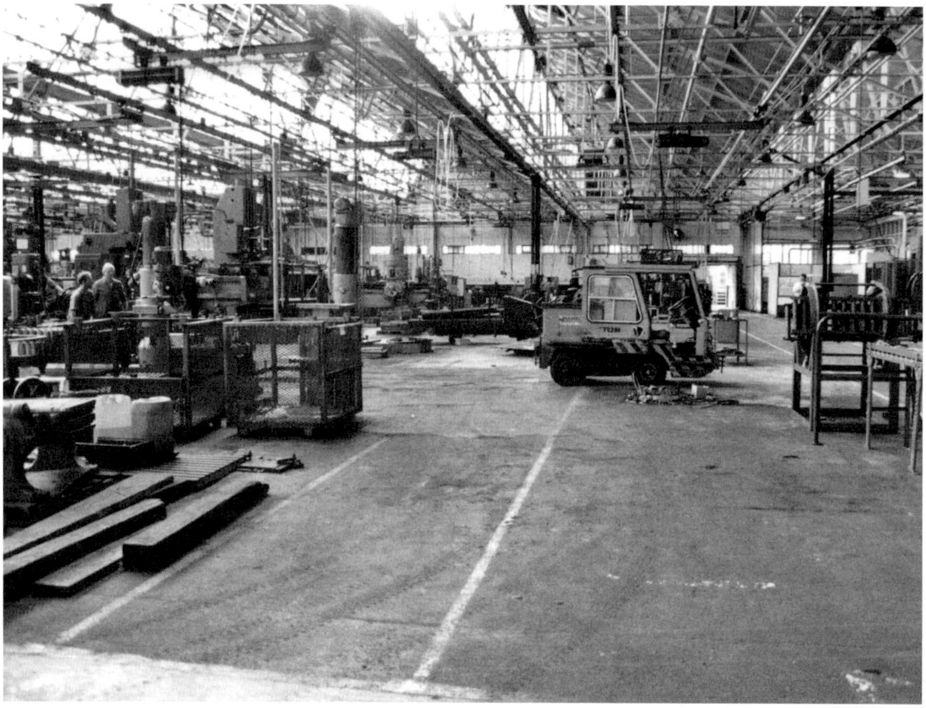

In 1997 there occurred a most dramatic development when RRMC stopped manufacturing engines at Pym's Lane. This view shows the machines being removed from the main shop.

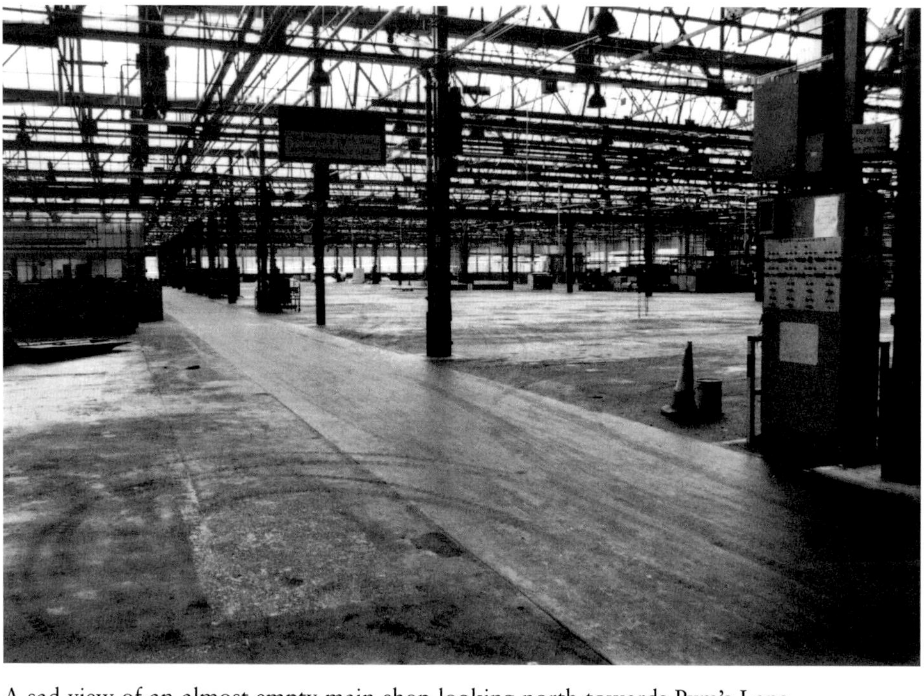

A sad view of an almost empty main shop looking north towards Pym's Lane.

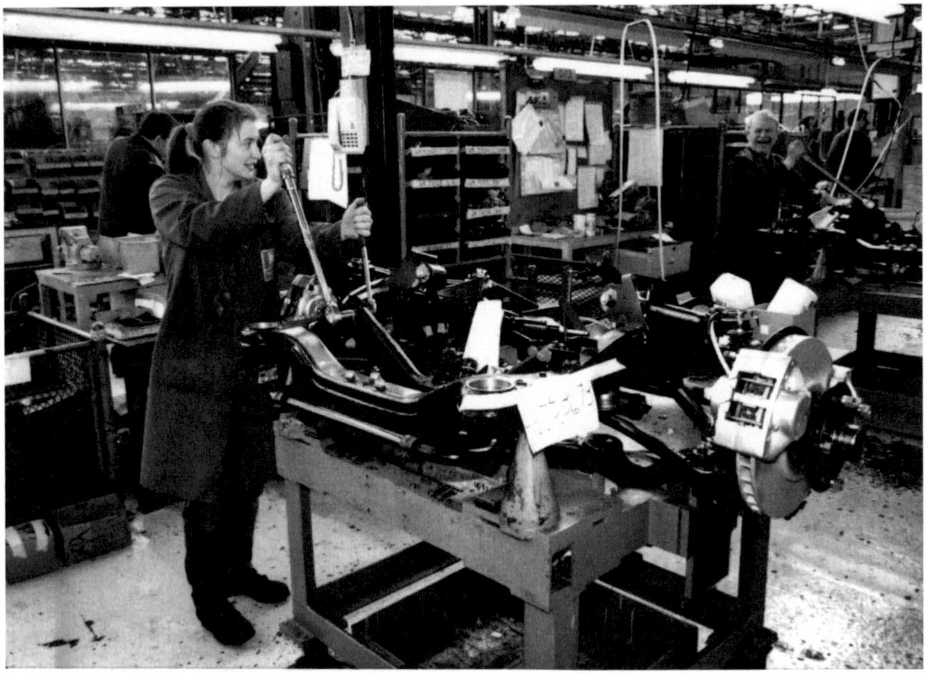

A sign of the increasing female incursion into previously male dominated jobs is a lady assembling a sub-frame.

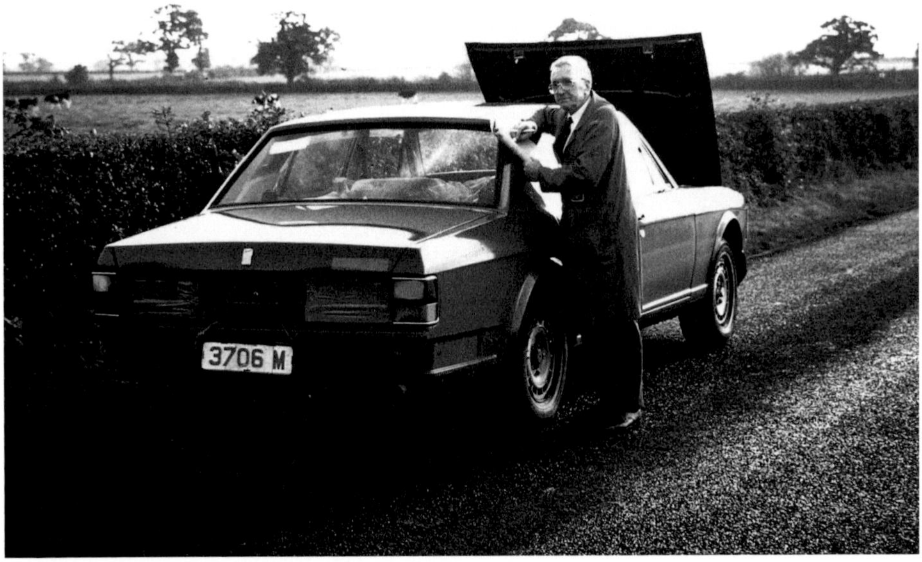

Maurice Whittaker road testing a Silver Spirit II on a Cheshire lane. Every car was road tested on three separate occasions.

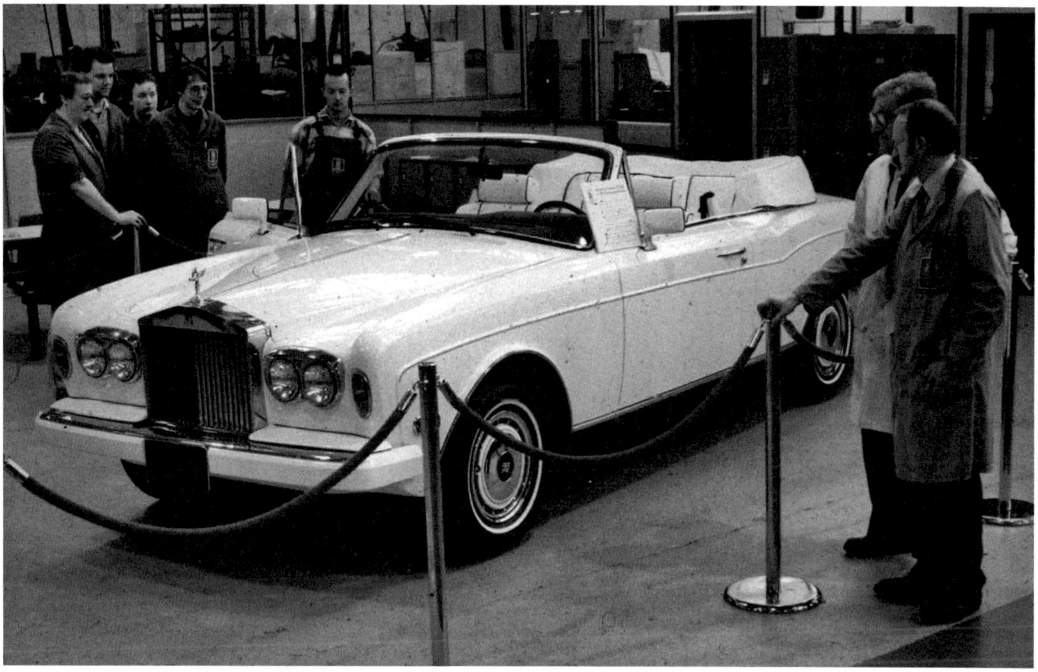

A Rolls-Royce Corniche displayed in the main shop being perused by Bob Tomkinson and R. Simms.

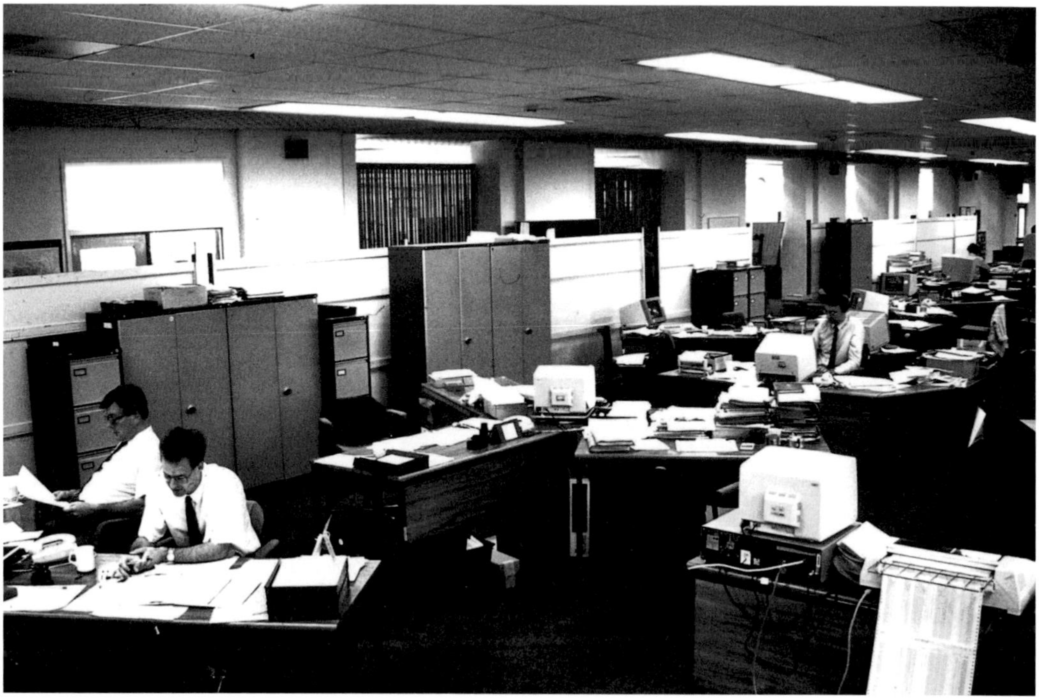

A general view of the purchase office which was responsible for buying in parts and other supplies. Angelo Ikin is nearest the camera.

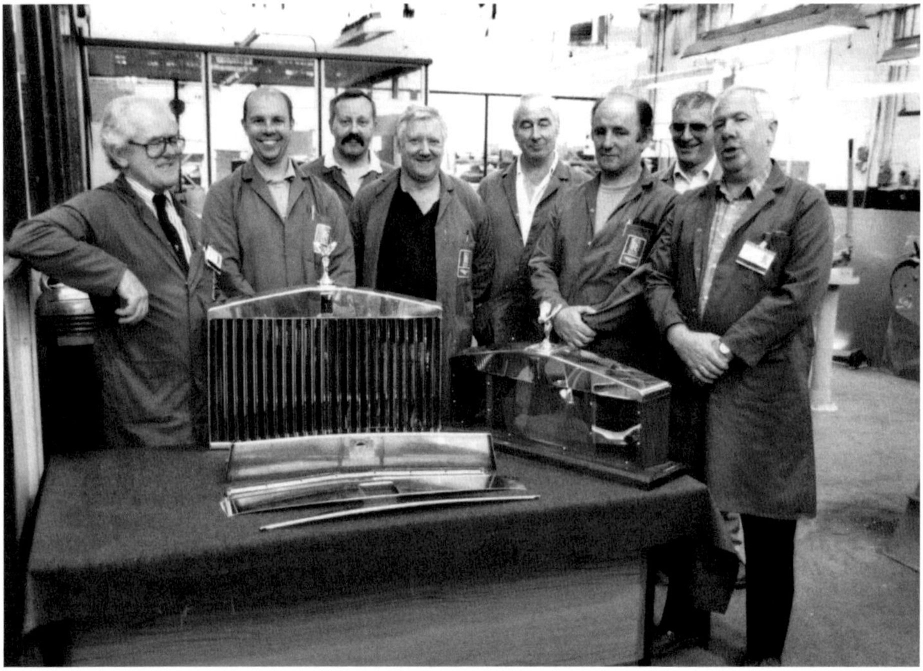

Craftsmen responsible for manufacturing the famous radiator with samples of their skill.

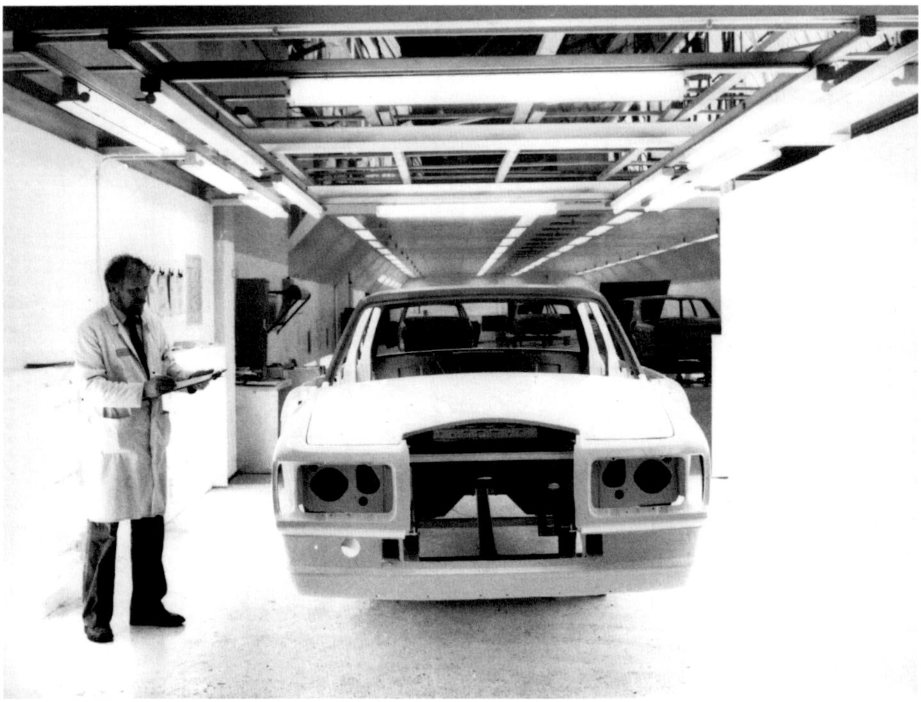

A Silver Spirit body, bought in from the Austin-Rover group, being inspected after paint (*c.* 1996).

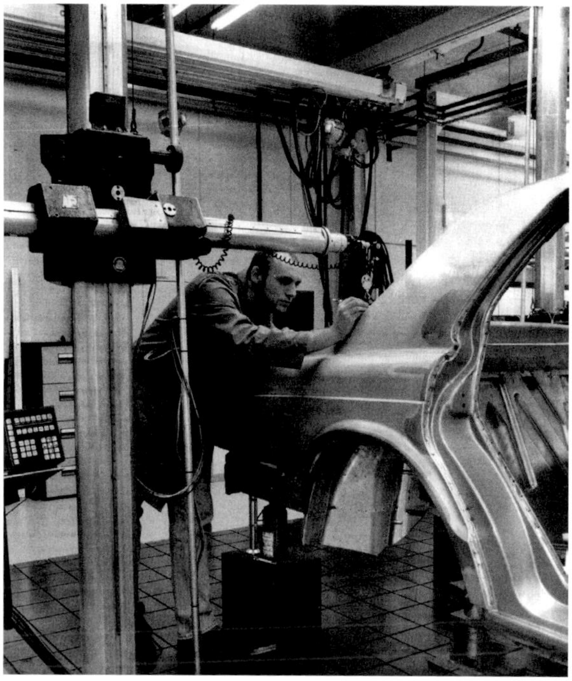

Right: Computer inspection of a 'bought in' body.

Below: In 1998 RR began to manufacture steel bodies 'in house'. In this image a machine is spot welding a Silver Seraph body.

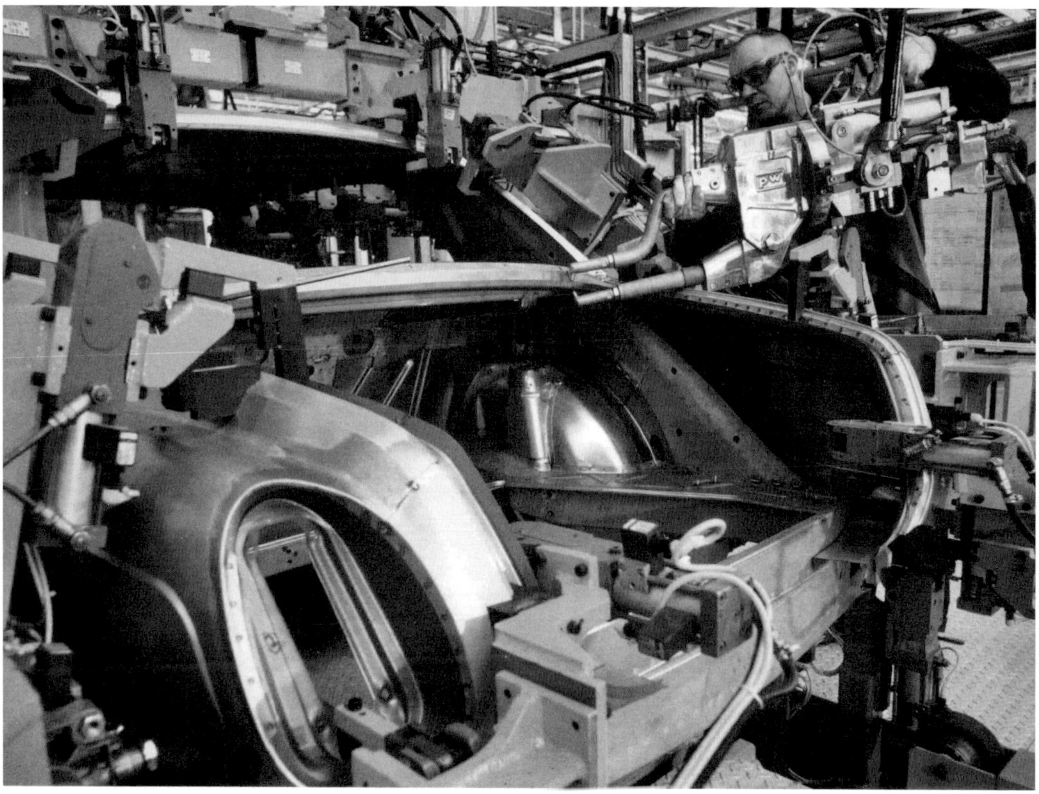

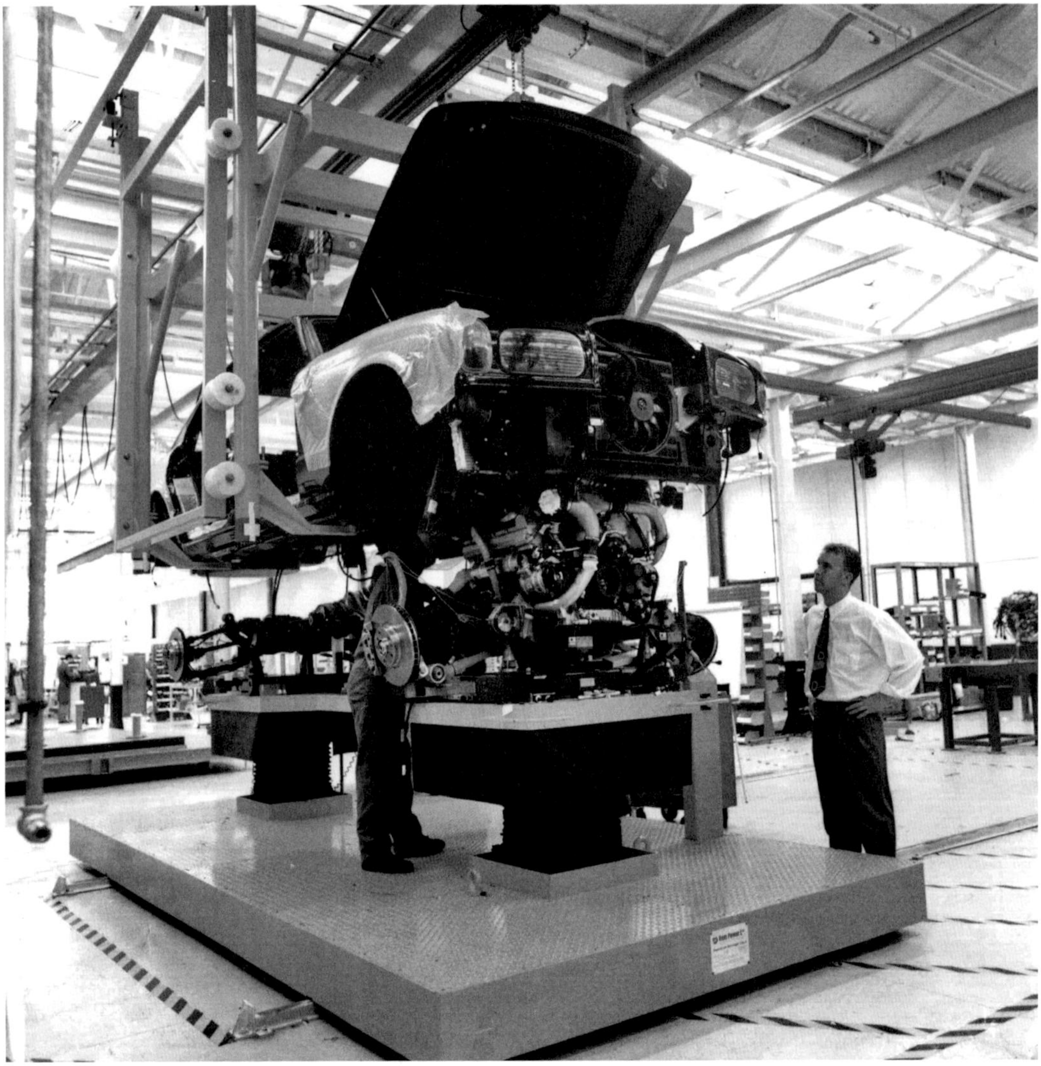

Neil Wilford watching a Bentley Arnage being mounted on to its sub-frames in the recently formed build hall, part of the massive changes that were being introduced at Pym's Lane *c.* 1998.

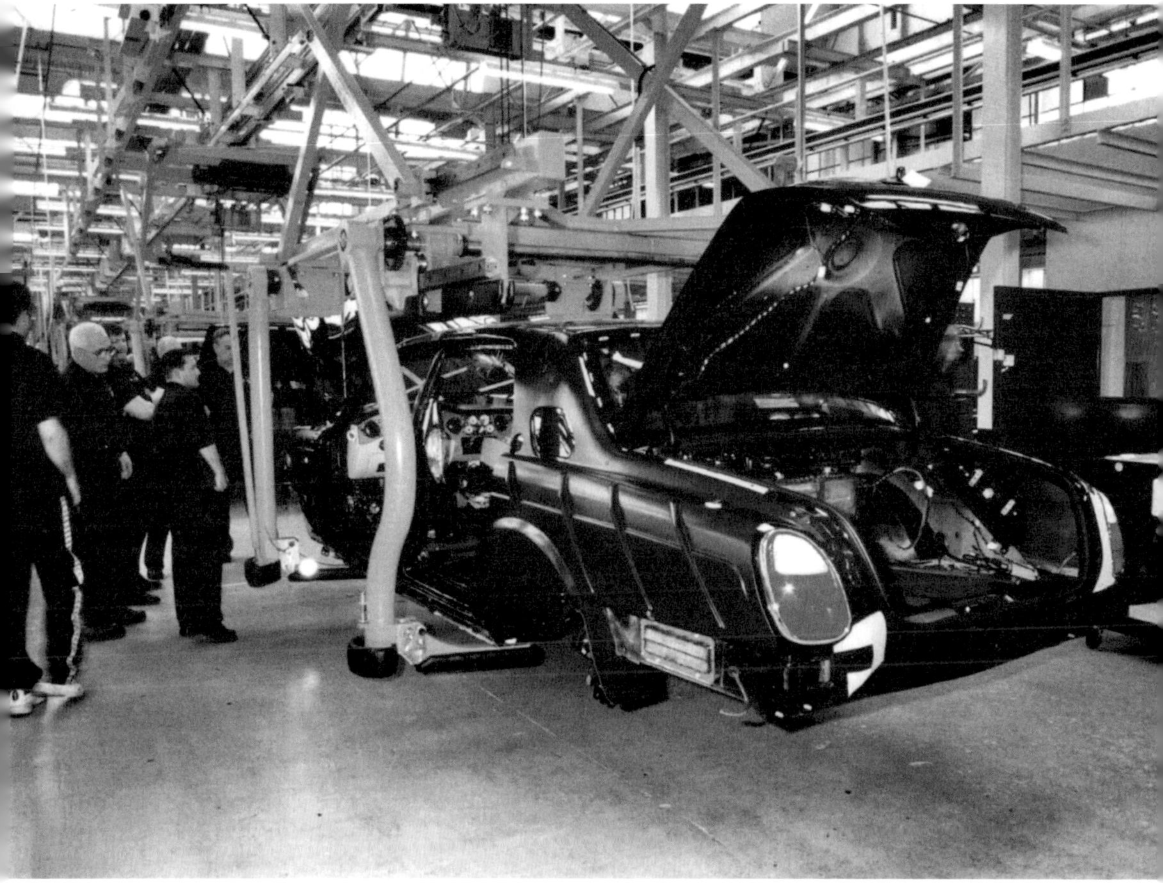

A Bentley Arnage in the build hall. Dave Preece, the last man to work for fifty years at Pym's Lane, is in the centre of the group.

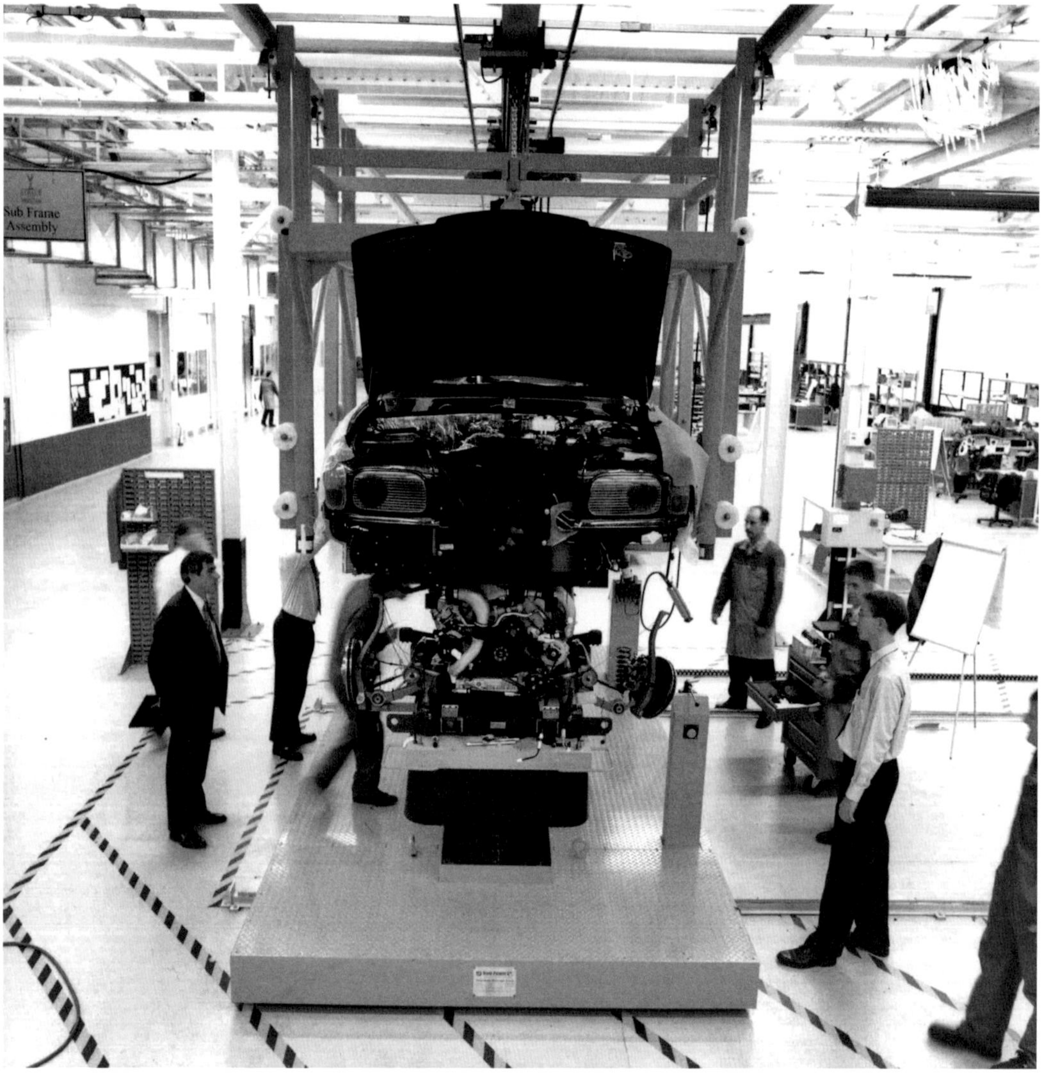

Russ Byford watches a pre-production Silver Seraph being married to the sub-frames in the build hall.

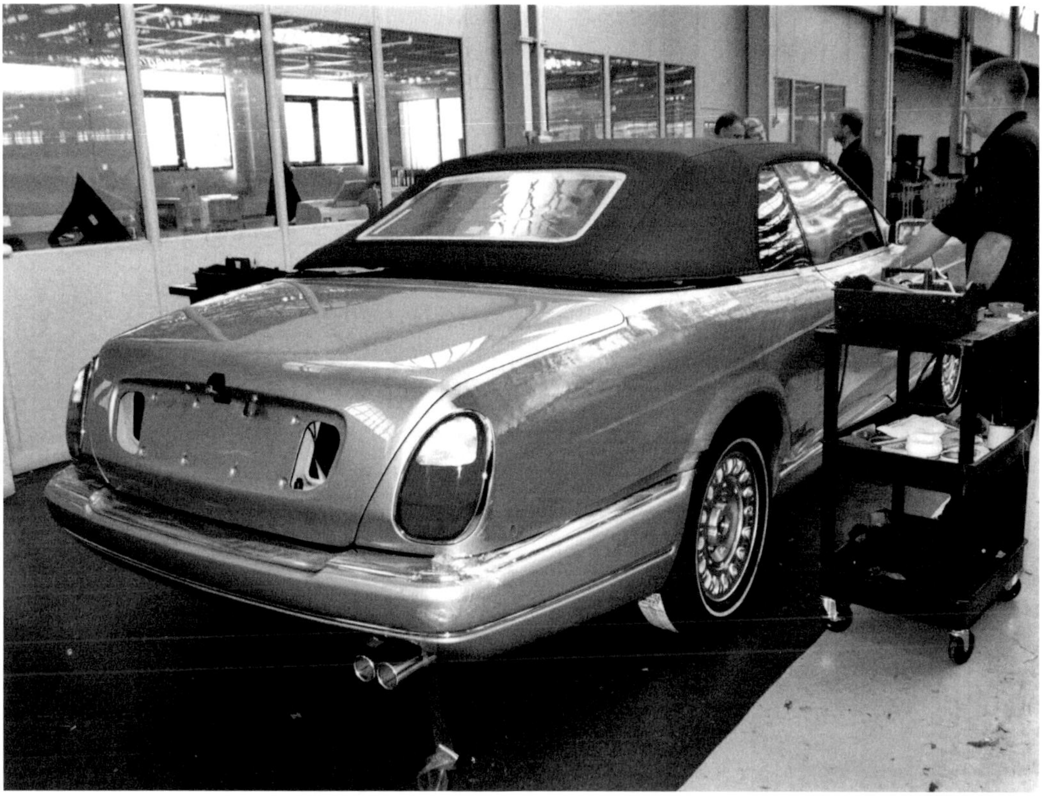

A Rolls-Royce Corniche after being assembled in 2002. In the background is the quality office. This particular range of Corniche cars came out in 2000 retaining the Rolls-Royce V8 engine originally designed by Jack Phillips.

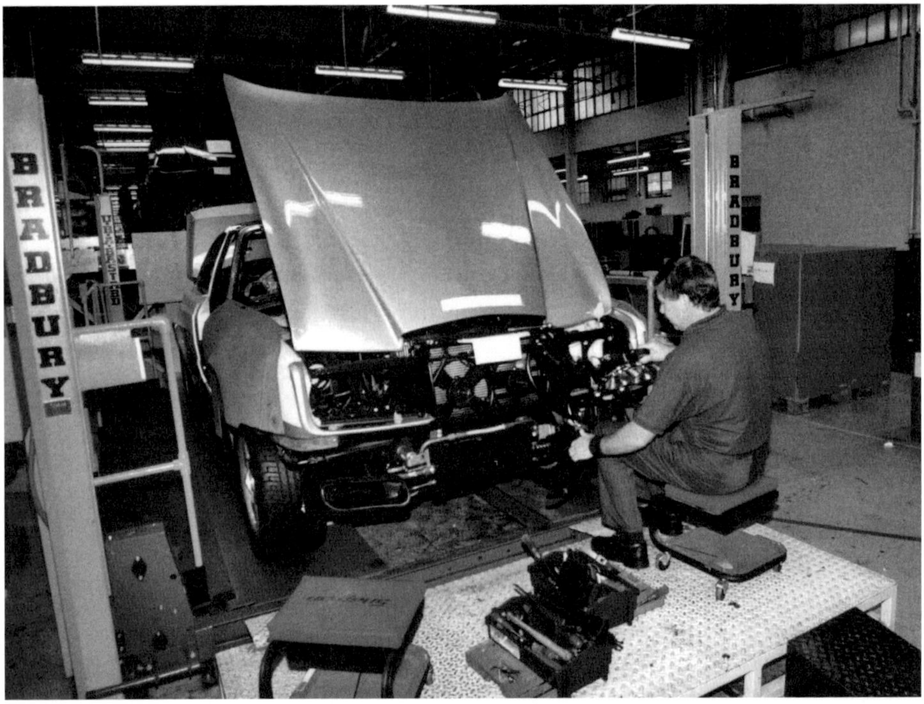

Fitting ancillaries to the front end of a Bentley Continental R.

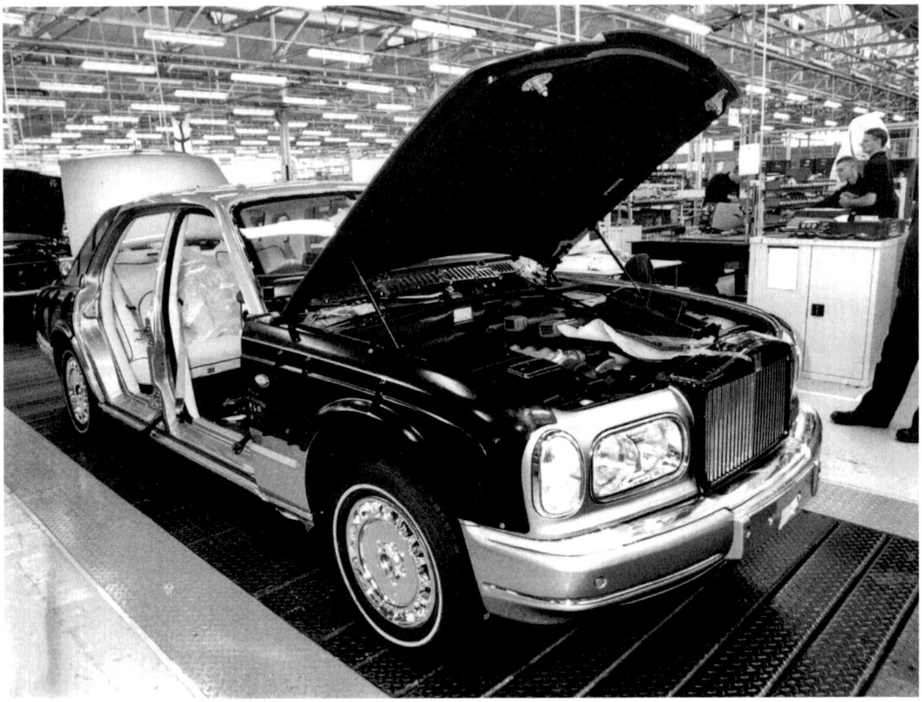

Last of the line Silver Seraph (2002) on the moving road in the build hall.

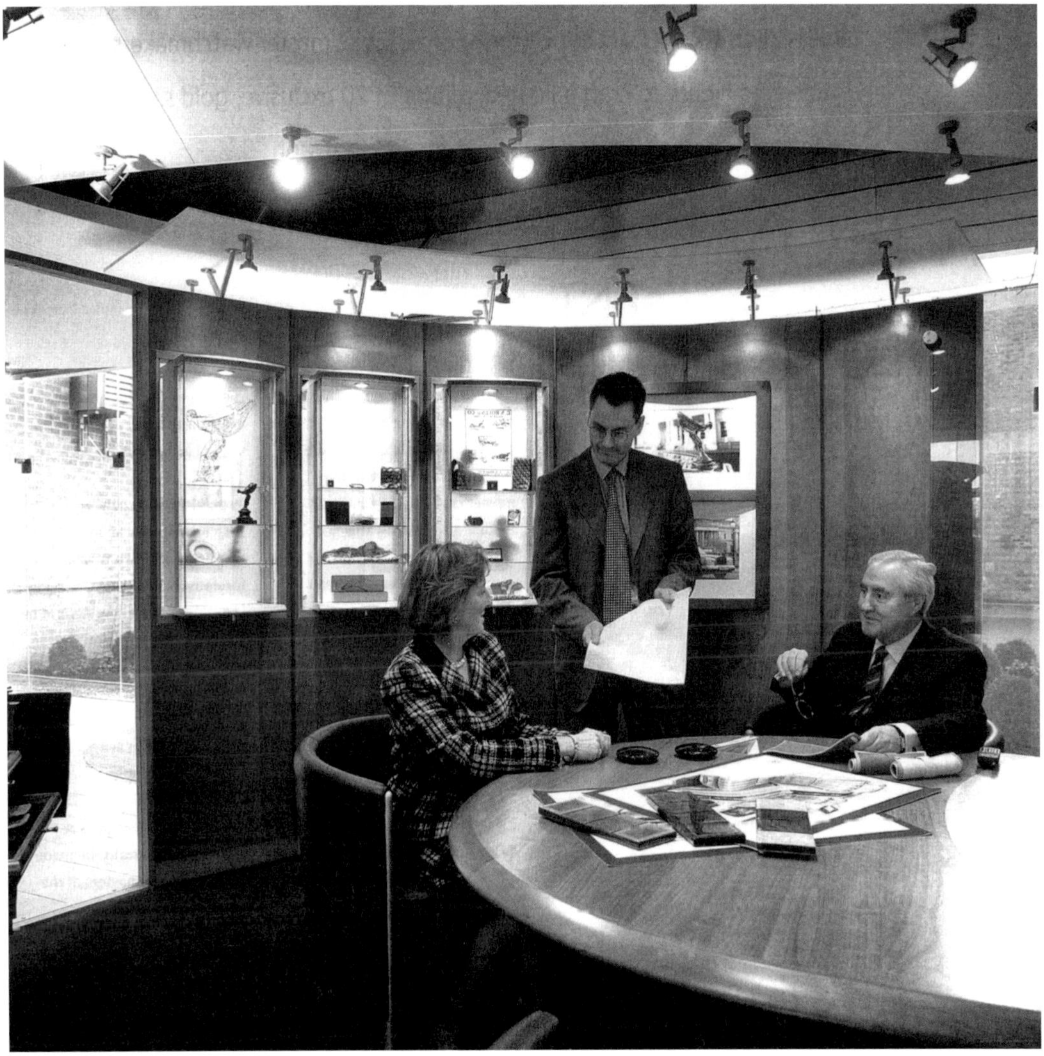

Another innovation of this era was the customer commission area where special requests from prospective owners were noted.

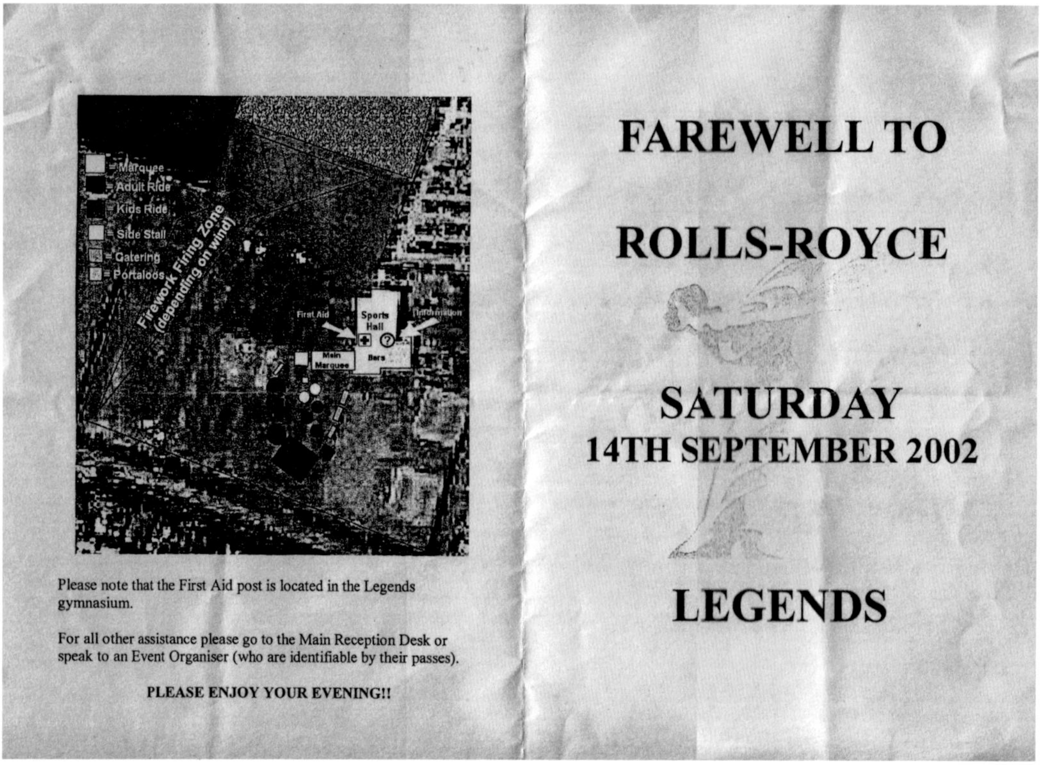

When Rolls-Royce left Crewe for pastures new in 2002 a programme of events was arranged to mark the occasion.

eight

PEOPLE & EVENTS 1938–1998

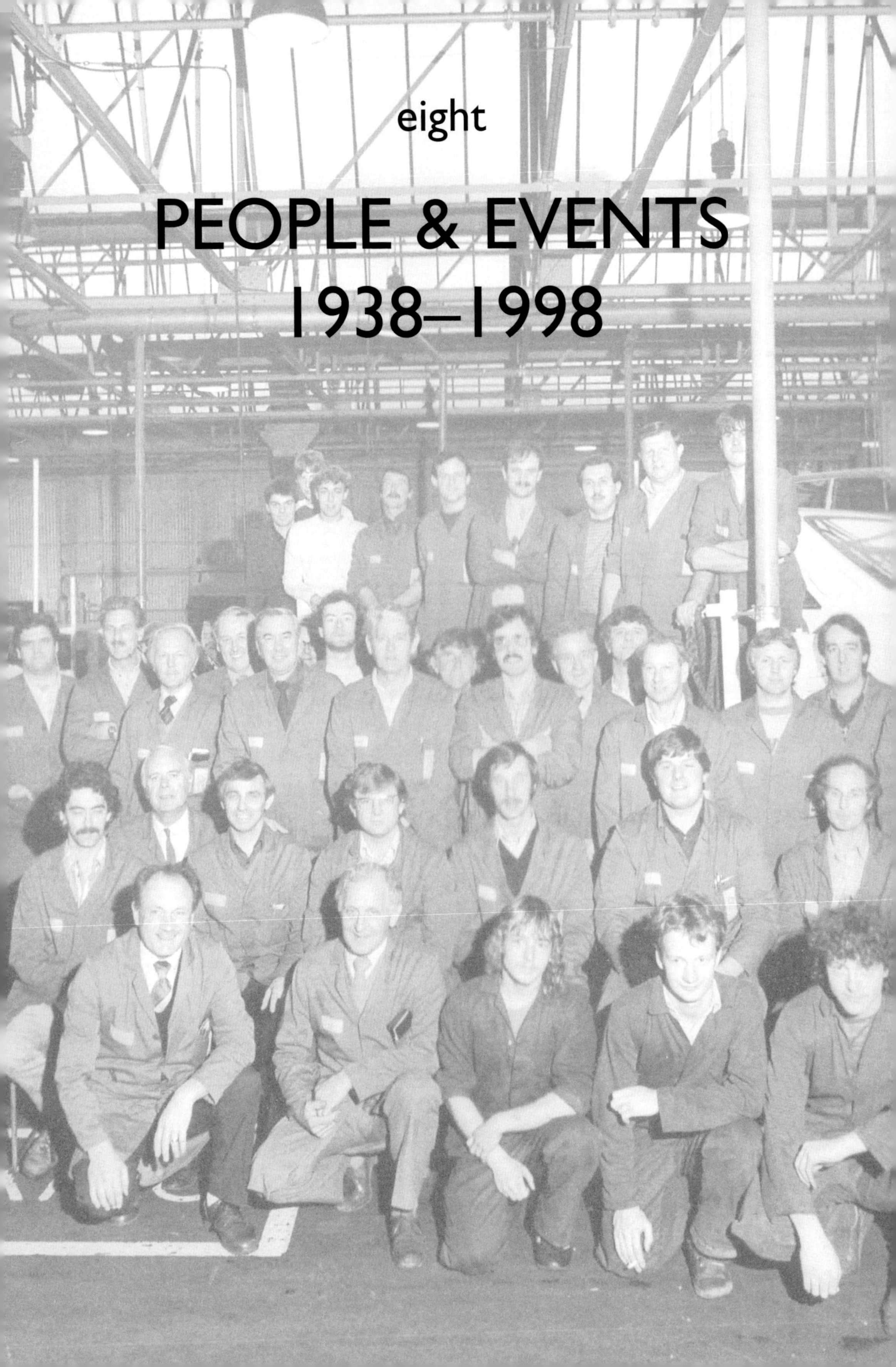

In this chapter we wish to illustrate some of the persons who worked at or who visited the factory. We also wish to demonstrate that RR offered to the employees sports, social and welfare facilities second to none.

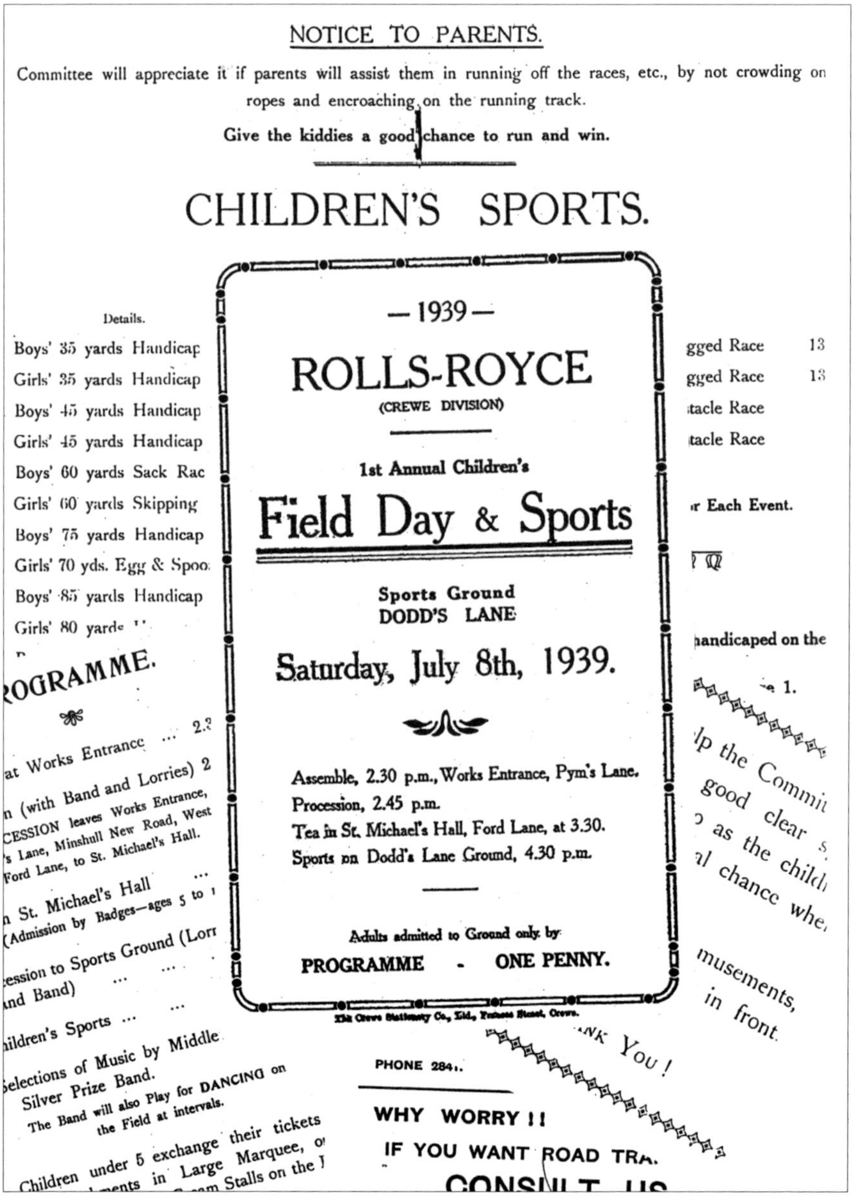
NOTICE TO PARENTS.

Committee will appreciate it if parents will assist them in running off the races, etc., by not crowding on ropes and encroaching on the running track.

Give the kiddies a good chance to run and win.

CHILDREN'S SPORTS.

— 1939 —

ROLLS-ROYCE

(CREWE DIVISION)

1st Annual Children's

Field Day & Sports

Sports Ground
DODD'S LANE

Saturday, July 8th, 1939.

Assemble, 2.30 p.m., Works Entrance, Pym's Lane.
Procession, 2.45 p.m.
Tea in St. Michael's Hall, Ford Lane, at 3.30.
Sports on Dodd's Lane Ground, 4.30 p.m.

Adults admitted to Ground only by
PROGRAMME - ONE PENNY.

Details.
Boys' 35 yards Handicap
Girls' 35 yards Handicap
Boys' 45 yards Handicap
Girls' 45 yards Handicap
Boys' 60 yards Sack Rac
Girls' 60 yards Skipping
Boys' 75 yards Handicap
Girls' 70 yds. Egg & Spoo
Boys' 85 yards Handicap
Girls' 80 yards

PHONE 284.

WHY WORRY !!
IF YOU WANT ROAD TRA

Despite the threat of war the welfare department organised, in 1939, the first of many field days for the children of employees.

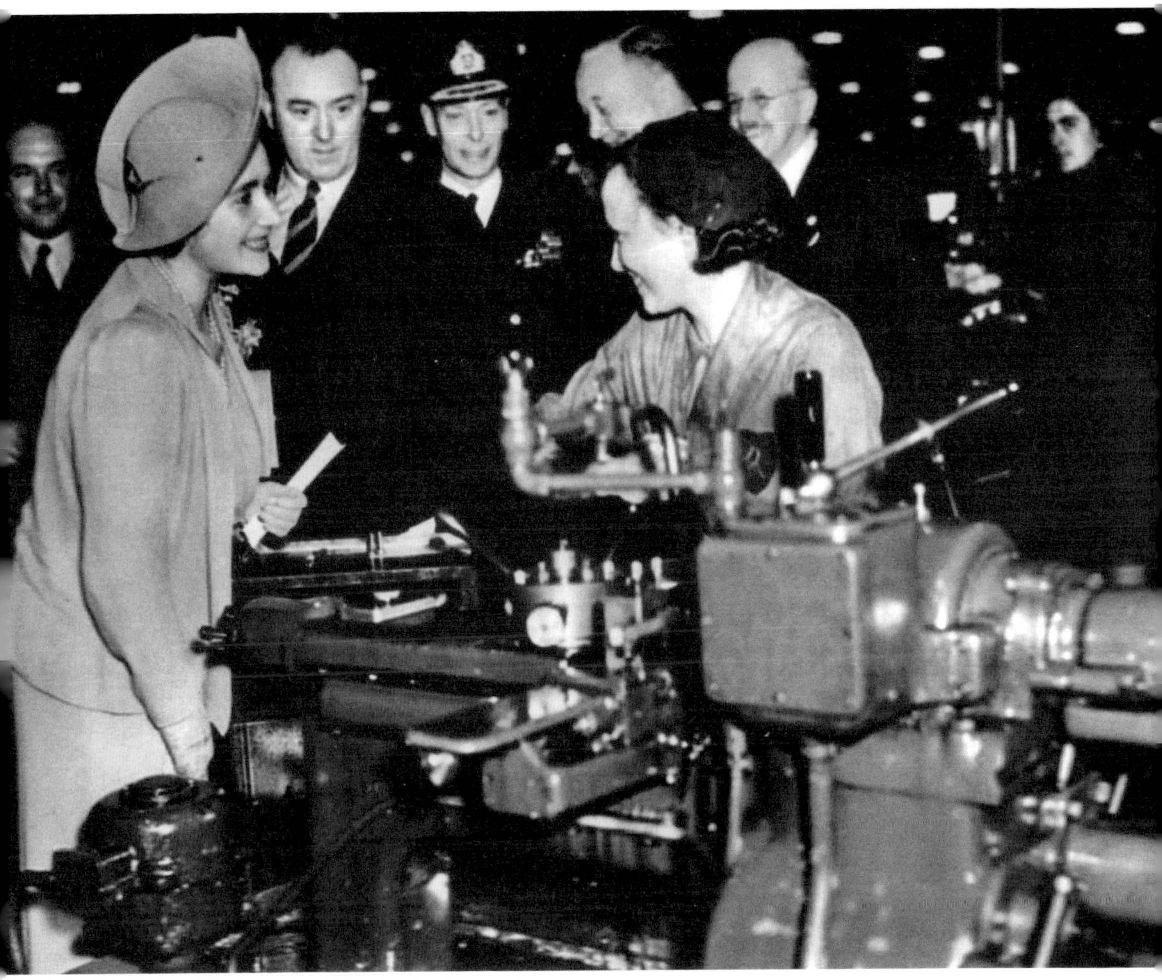

George VI and Queen Elizabeth on a visit to Pym's Lane. John Morris, general manager, is alongside her majesty. Arthur Sidgreaves, MD of Rolls-Royce is behind the capstan operator.

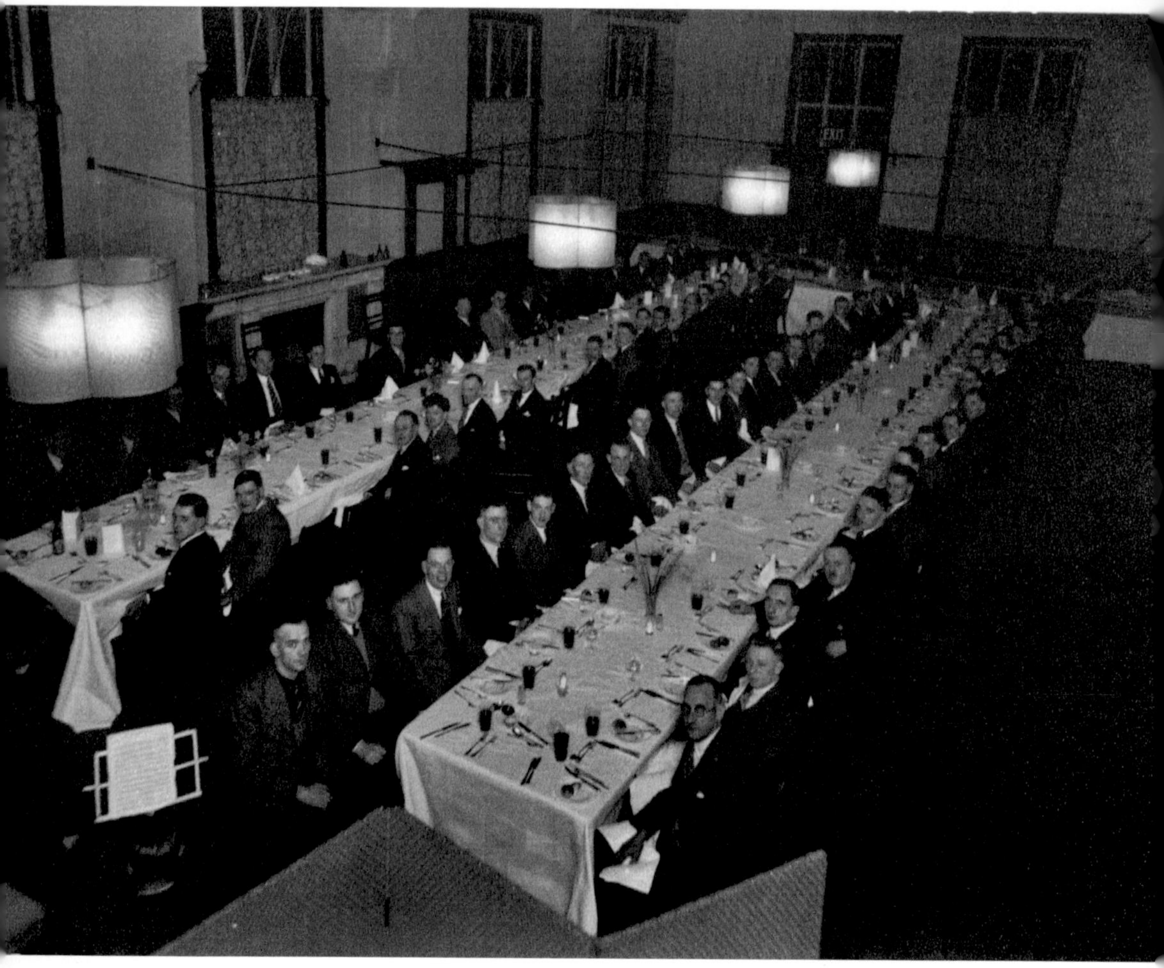

Test department dinner in 1940. From the number present there could only have been a skeleton staff on duty.

Right: Some of the documents generated by the official enquiry into the bombing of December 1940.

Below: Dust covers over the machines in the bombed machine shop.

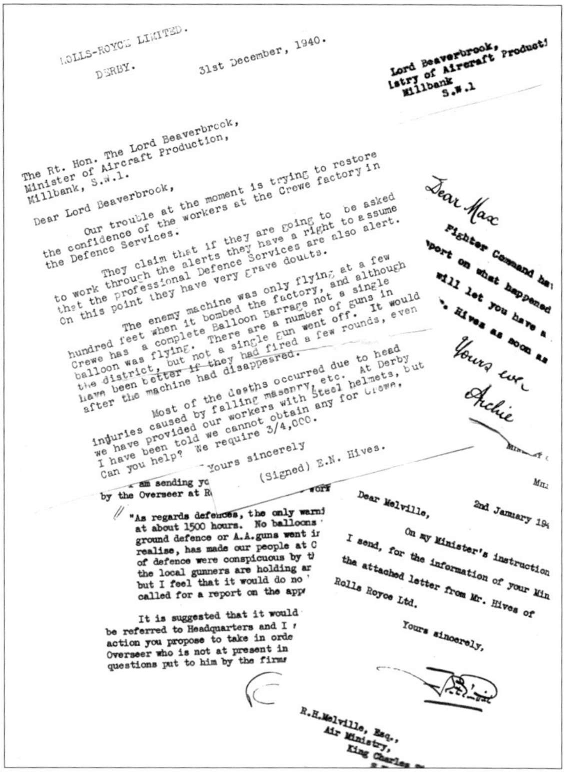

ROLLS-ROYCE LIMITED.
DERBY. 31st December, 1940.

The Rt. Hon. The Lord Beaverbrook,
Minister of Aircraft Production,
Millbank, S.W.1.

Dear Lord Beaverbrook,

Our trouble at the moment is trying to restore the confidence of the workers at the Crewe factory in the Defence Services.

They claim that if they are going to be asked to work through the alerts they have a right to assume that the professional Defence Services are also alert. On this point they have very grave doubts.

The enemy machine was only flying at a few hundred feet when it bombed the factory, and although Crewe has a complete Balloon Barrage not a single balloon was flying. There are a number of guns in the district, but not a single gun went off. It would have been better if they had fired a few rounds, even after the machine had disappeared.

Most of the deaths occurred due to head injuries caused by falling masonry, etc. At Derby we have provided our workers with steel helmets, but I have been told we cannot obtain any for Crewe. Can you help? We require 3/4,000.

Yours sincerely

(Signed) E.N. Hives.

Lord Beaverbrook,
istry of Aircraft Product
Millbank
S.W.1

Dear Max

Fighter Command ha
port on what happened
will let you have a
. Hives as soon as

Yours ever
Archie

I am sending yo
by the Overseer at R

"As regards defences, the only warn
at about 1500 hours. No balloons
ground defence or A.A.guns went i
realise, has made our people at C
of defence were conspicuous by t
the local gunners are holding a
but I feel that it would do no
called for a report on the app

It is suggested that it would
be referred to Headquarters and I
action you propose to take in orde
Overseer who is not at present in
questions put to him by the firm

Dear Melville, 2nd January 194

On my Minister's instruction I send, for the information of your Min the attached letter from Mr. Hives of Rolls Royce Ltd.

Yours sincerely,

R.H.Melville, Esq.,
Air Ministry,
King Charles

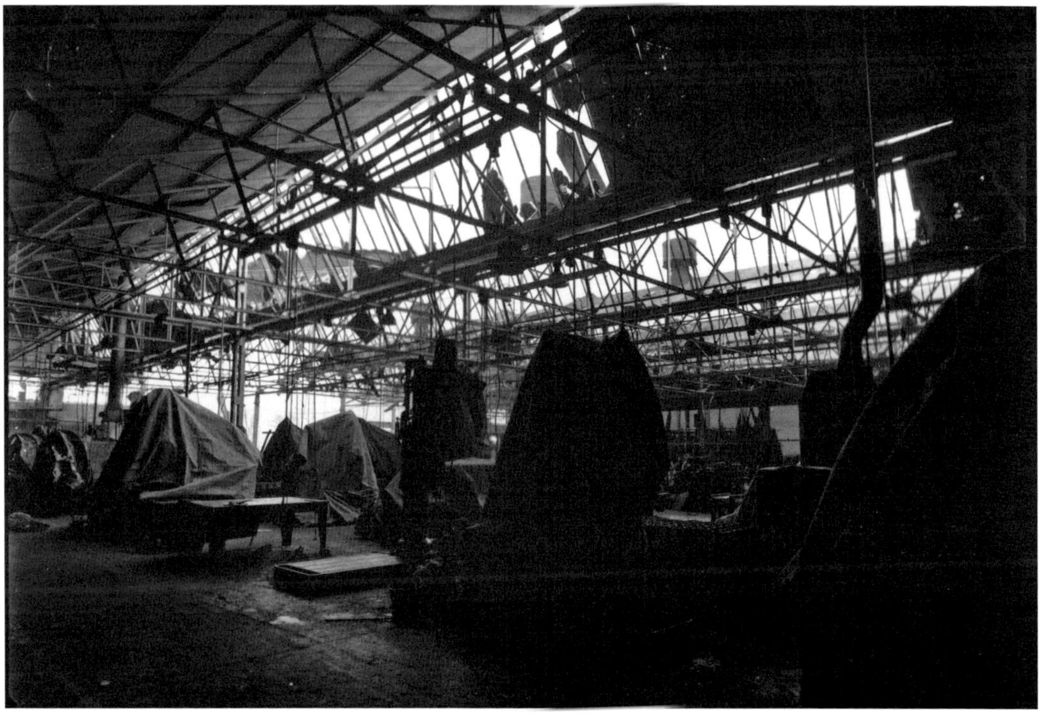

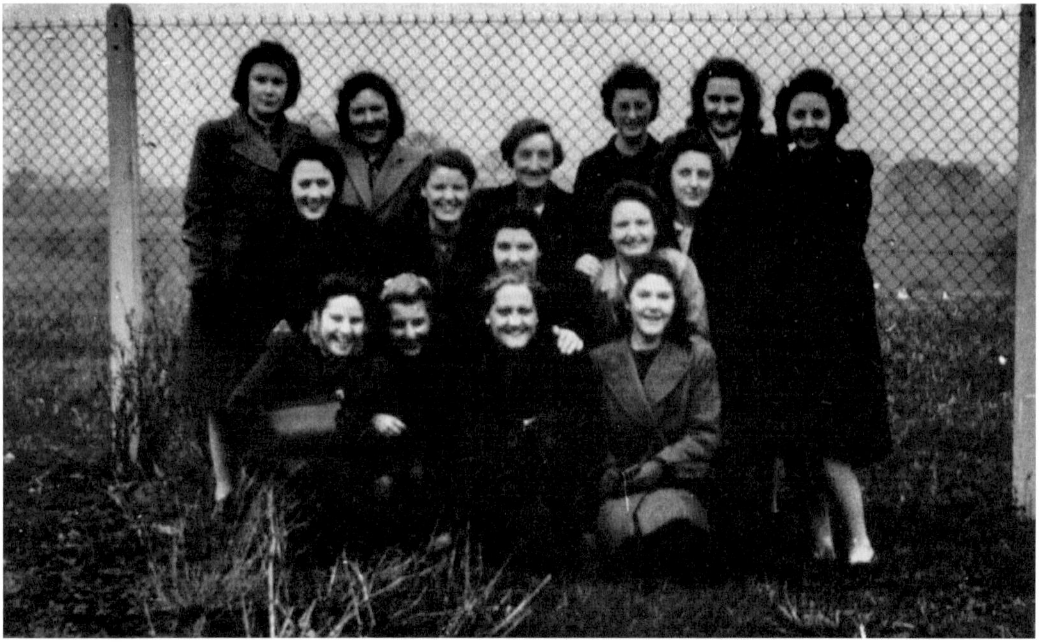

The ladies of the typing bureau in 1944. Miss Thompson, one of the first employees from August 1938 is in the middle of the back row. Some of the others are F. Palin, D. Pole, D. Ollerhead, I. Egerton, J. Carnwell and M. Cooper.

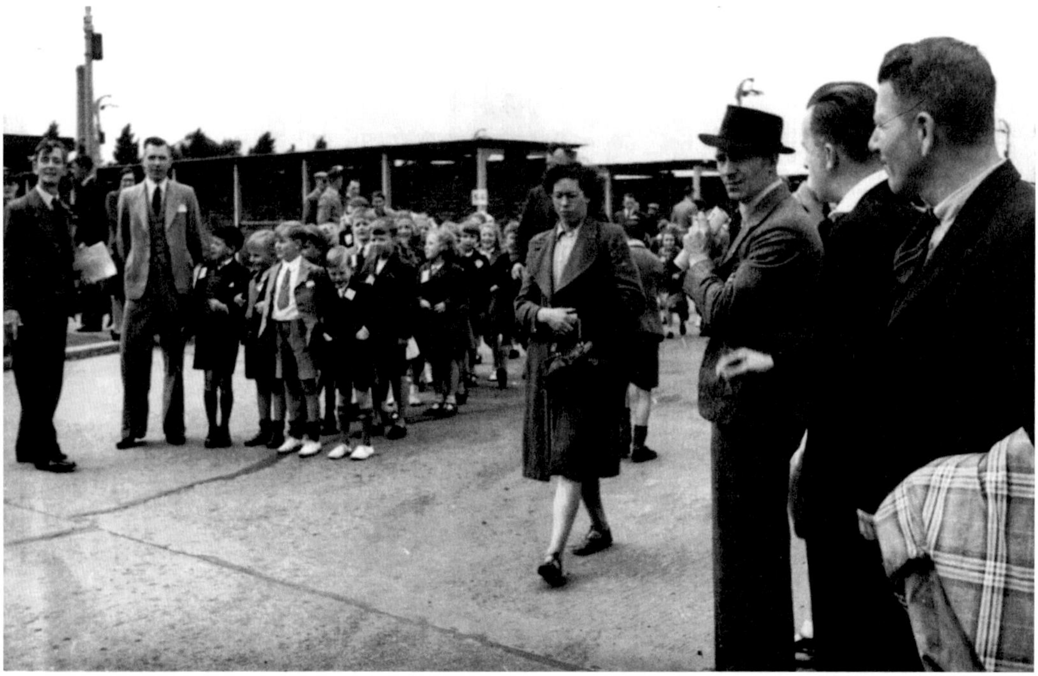

Children assembling for the field day in Pym's Lane car park (*c.* 1948). E. Hunstone, welfare manager, is holding a sheaf of papers.

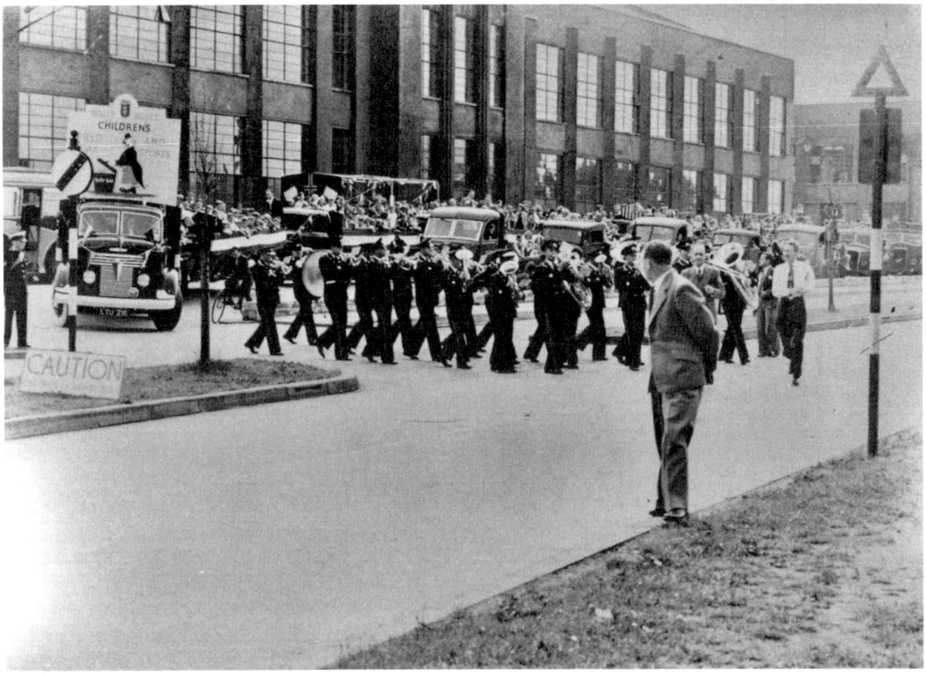

Procession to the sports ground for the field day (*c.* 1950). Bill Edge, personnel and Charles Elson, union convener, are immediately in front of the band.

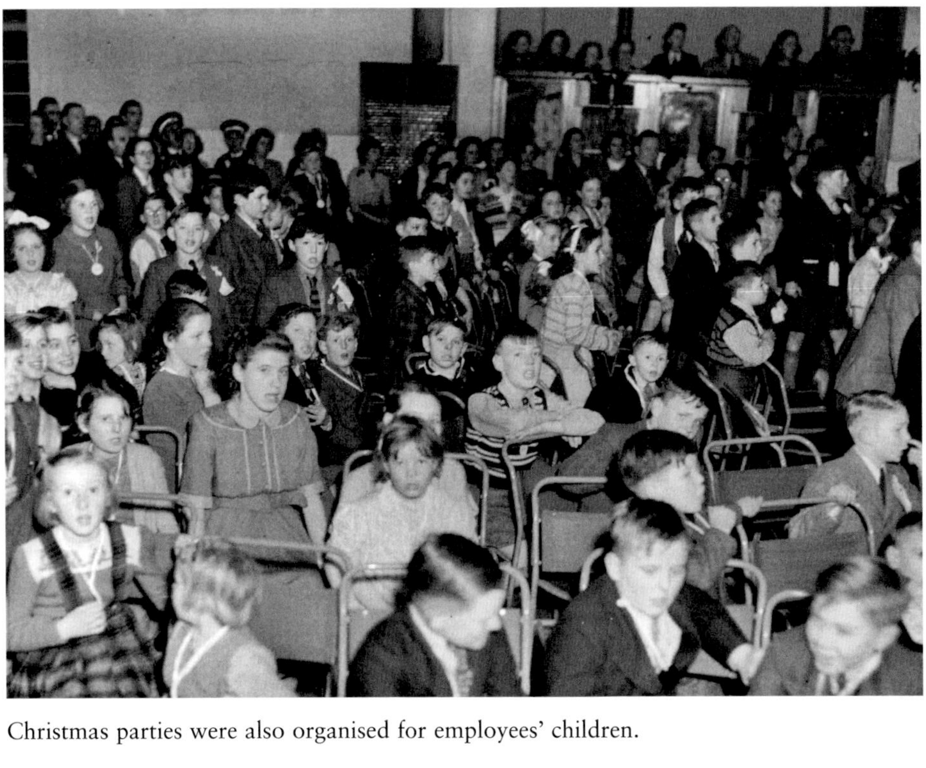

Christmas parties were also organised for employees' children.

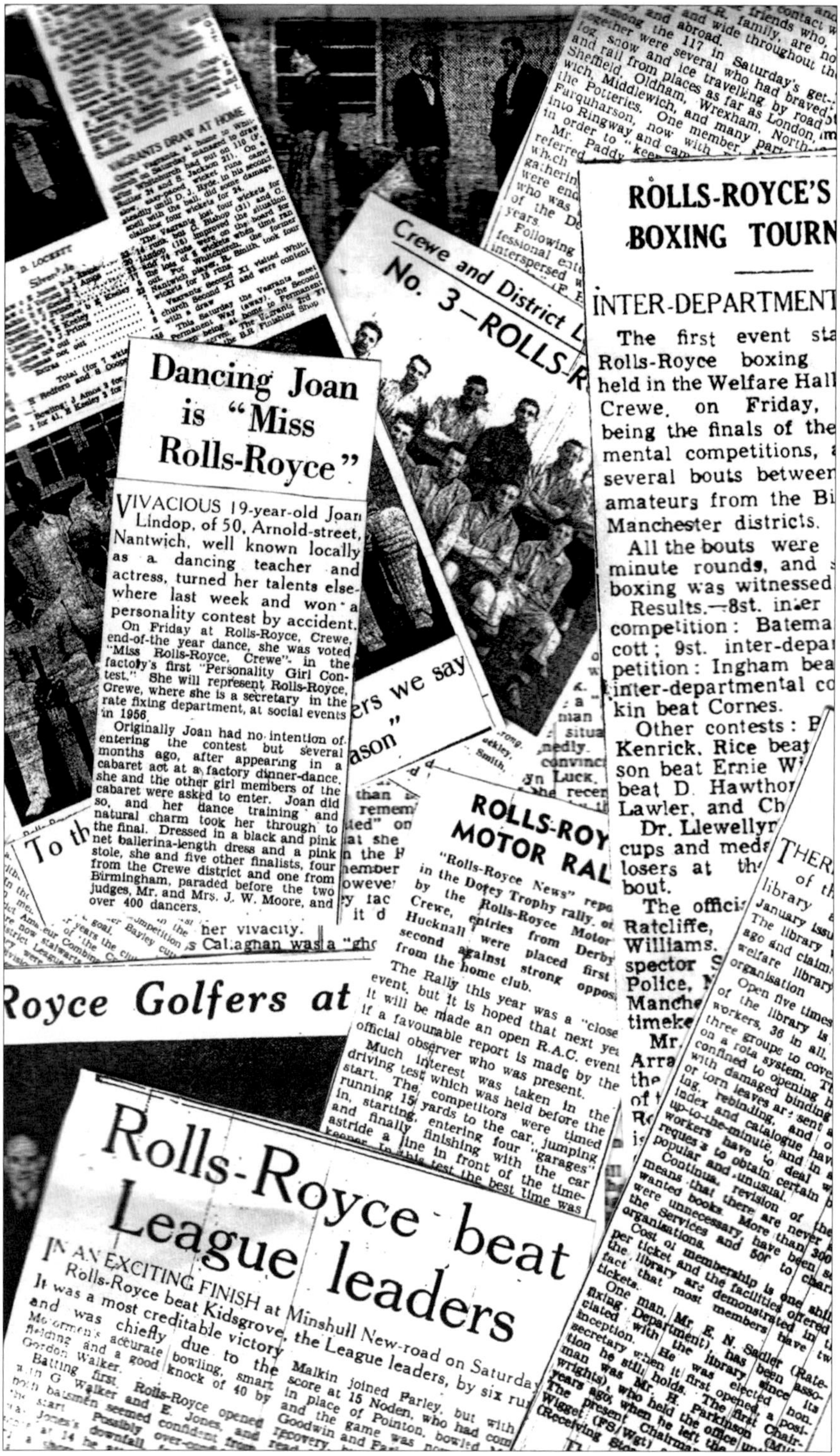

News of some of the activities of the welfare department reached the local paper.

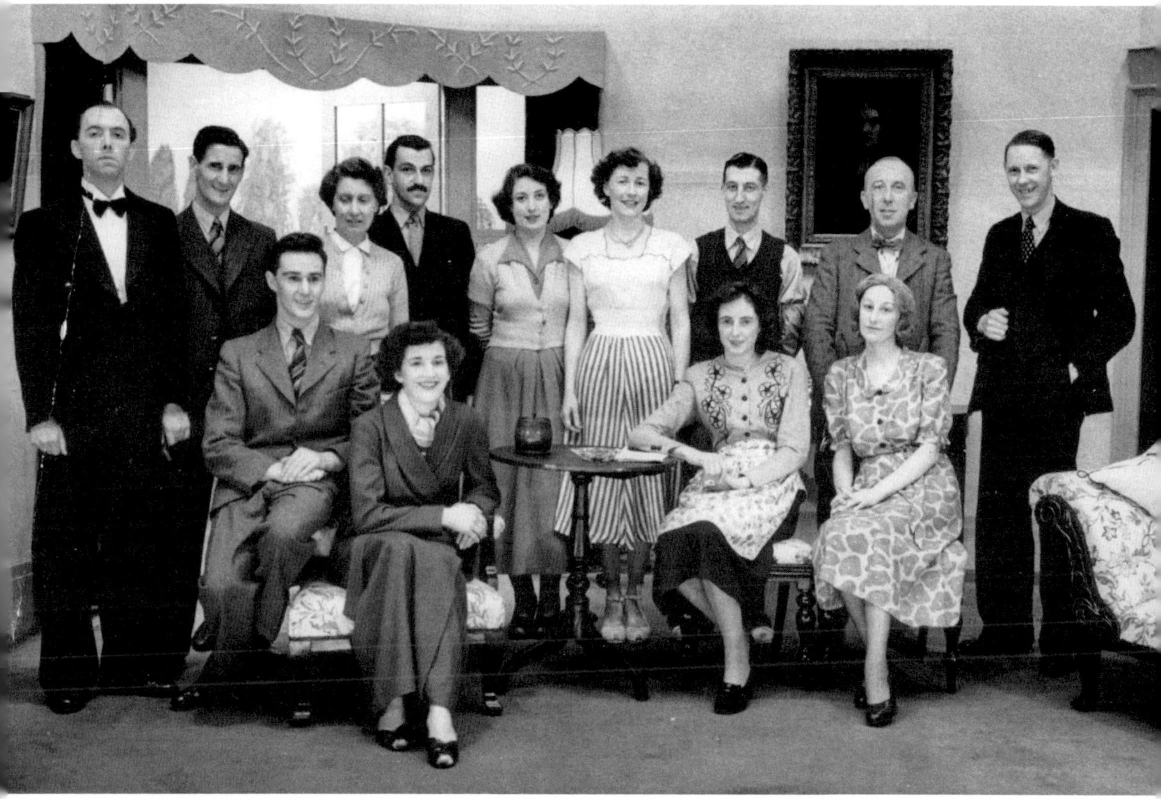

Amongst the cast of the drama section's production of *The Chiltern Hundreds* (c. 1950) are Arthur George, Peter Cantrill, Elsie Brown and Ellen Callaghan.

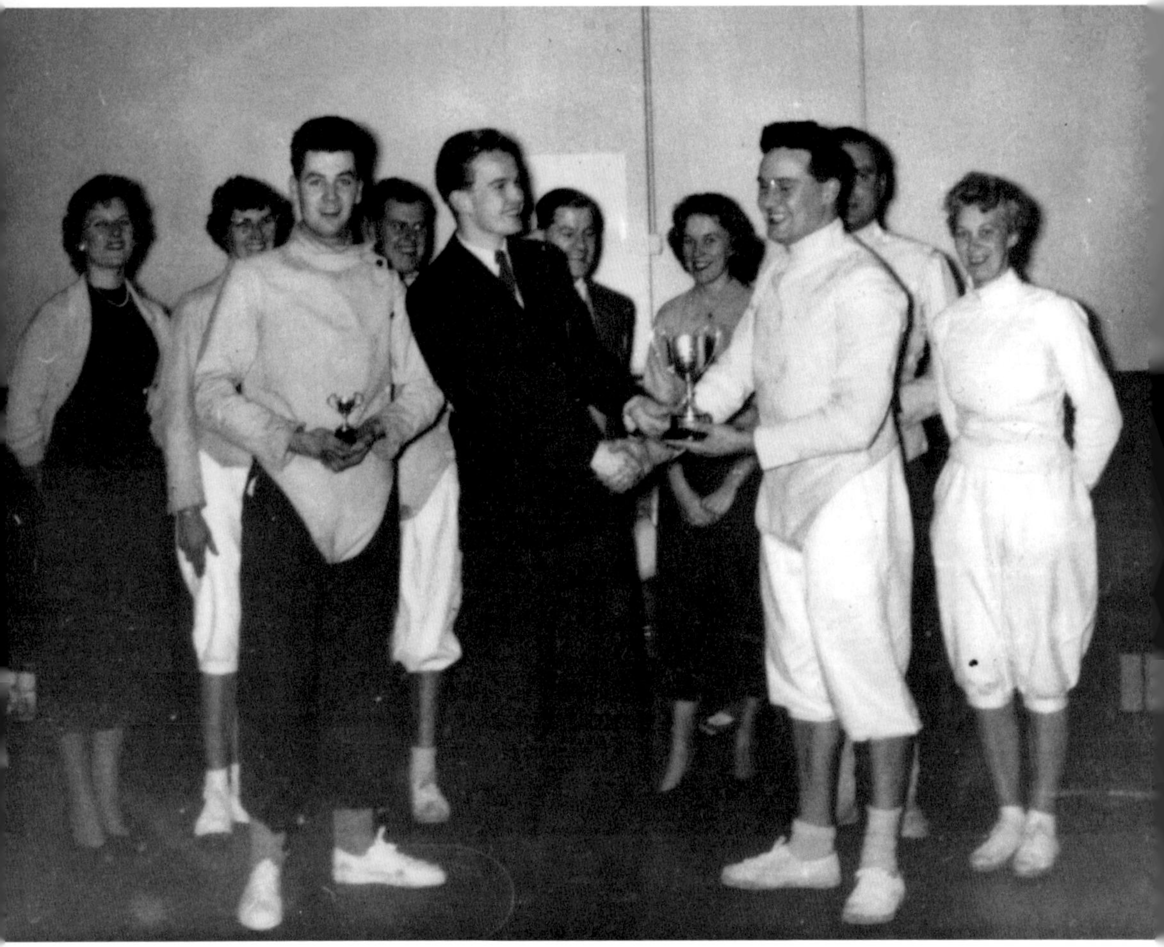

For a short while in the fifties a fencing section was active. Ron Ashley (manufacturing director 1979-85) is holding a trophy while Arthur George (jig and tool draughtsman) is receiving his.

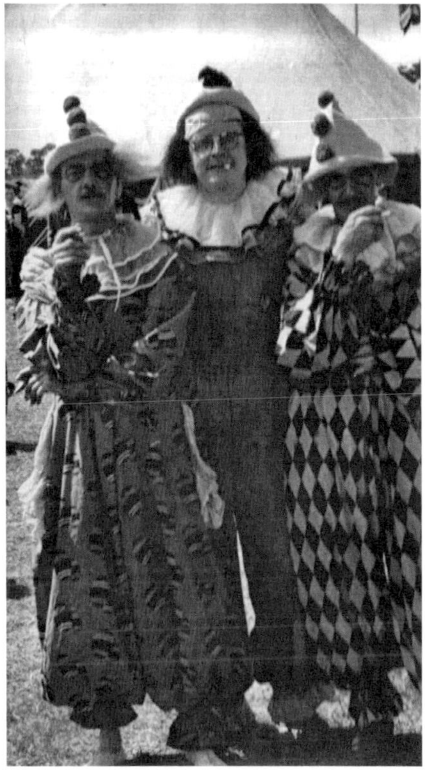

Right: Frank and Ted Ollerhead, with Sid Harvey in the centre, part of the Crazy Gang of company employees who entertained the children at the early field days.

Below: Facing the camera at a 1953 dinner are Bill Elliott (millwrights), N. Wiggett (ratefixers), W. Elliott, (equipment), H. Peak (chief draughtsman), unknown, V. Lewis (experimental). Ron Dyson is half turned with Geoff Bastow (development) in profile.

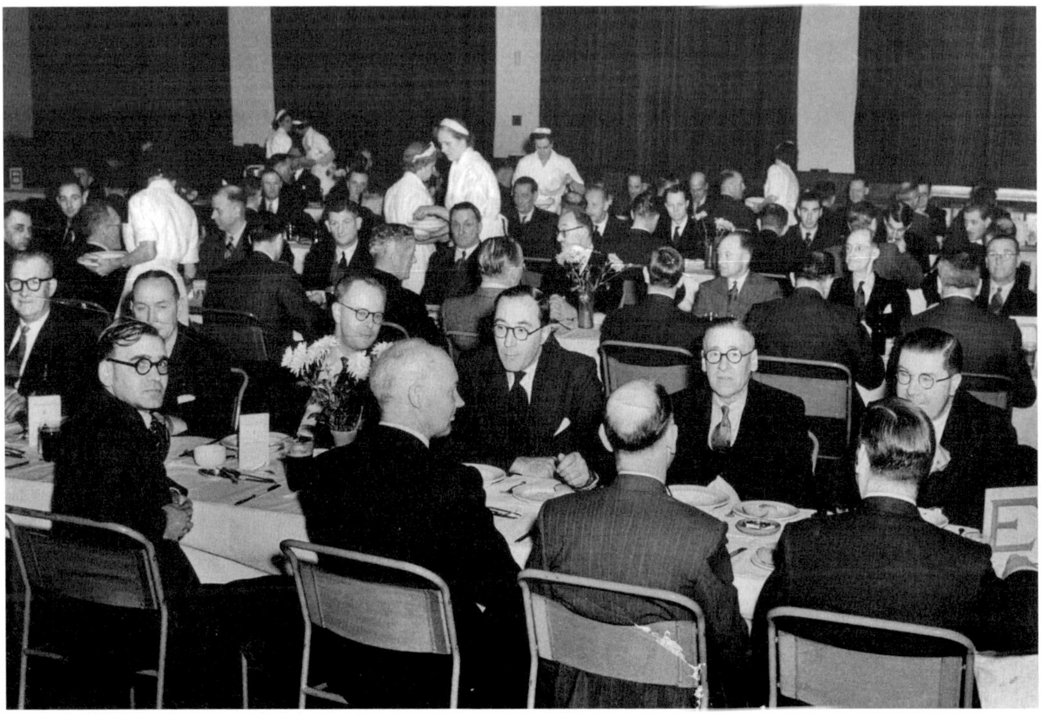

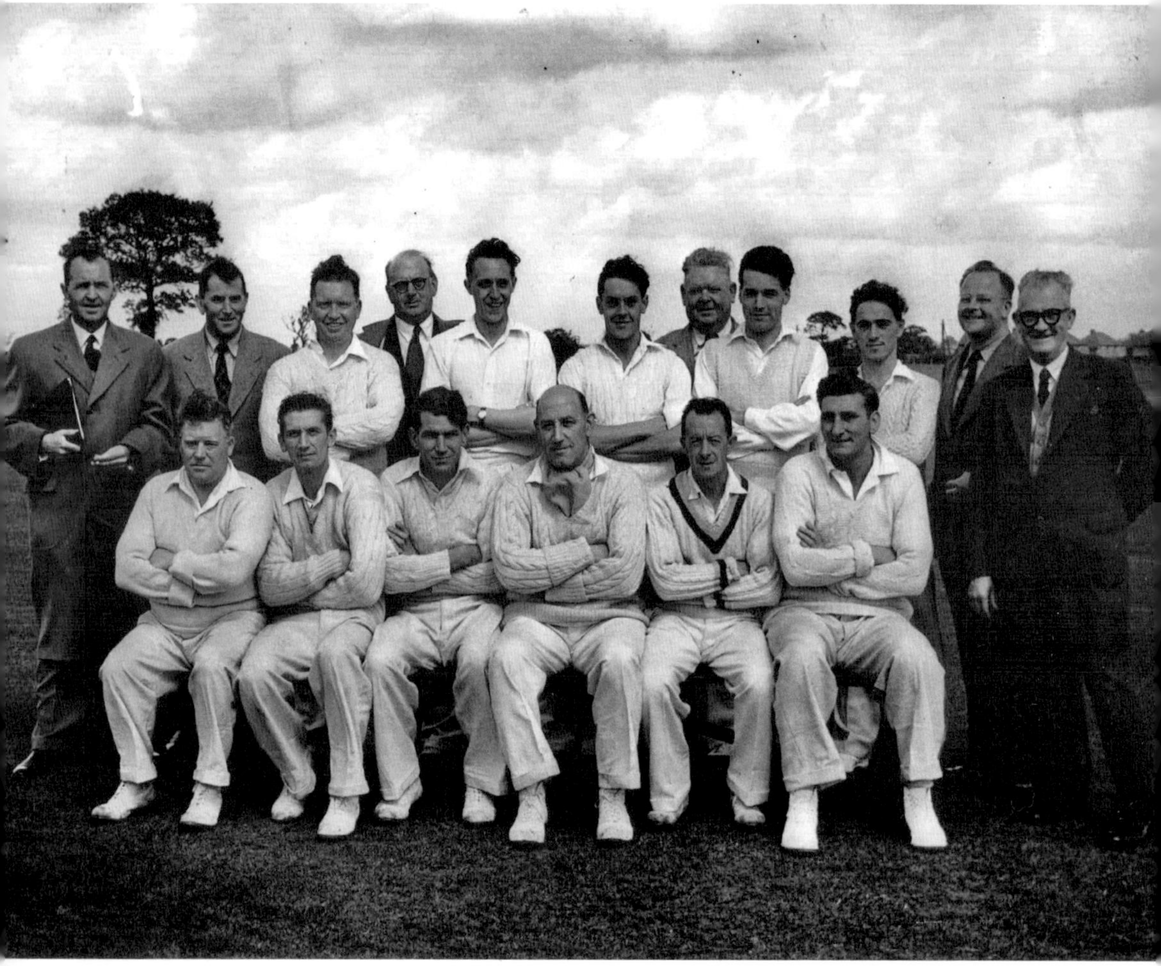

RR cricket team 1954. Back row: T. Byrne, J. Hewitson, J. Cotton, G. Perrins, E. Pointon, W. Bailey, L. Ord, M. Greenwood, C. Jones, W. Elliott, Bill Elliott. Front row: G. Walker, E. Jones, Jock Knight, J. Prince, E. Buchan, J. Amos. This sportsfield is now a housing estate.

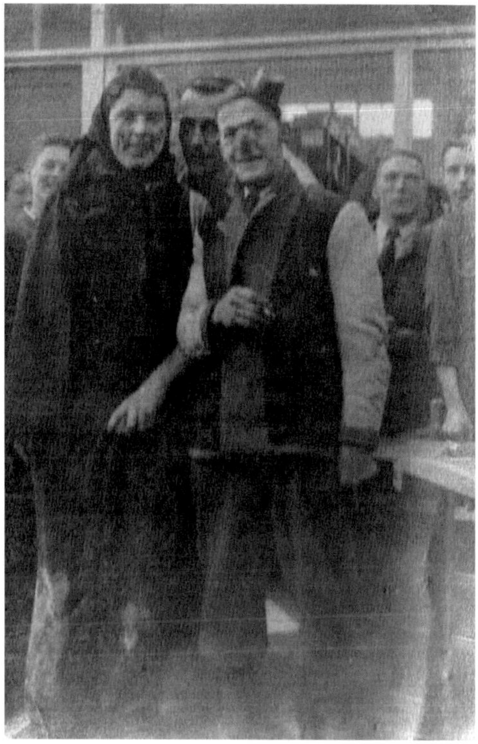

Right: In the forties and fifties a tradition appeared of apprentices' bun-fights at Christmas. Frank Manley of the camshaft line is ready to participate with his setter (author's father) as his second. Looking over Frank's shoulder is John Dobson, one of six brothers who worked at Pym's Lane.

Below: Post bun-fight experimental department apprentices (including Ken Wade and Bernard Preston).

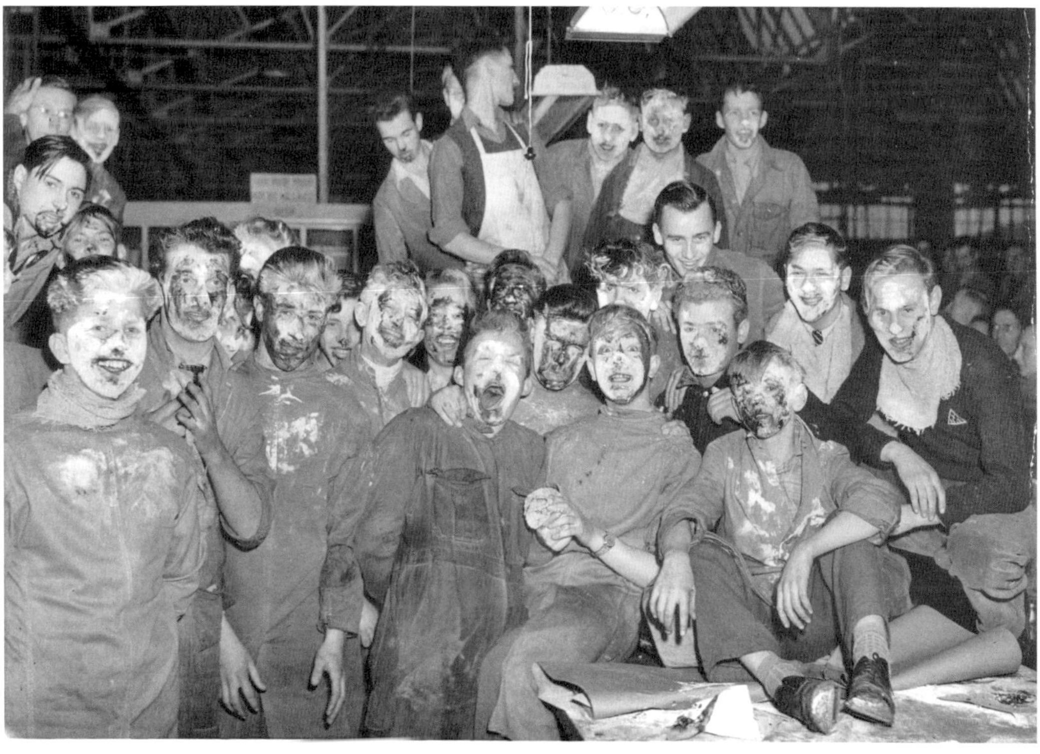

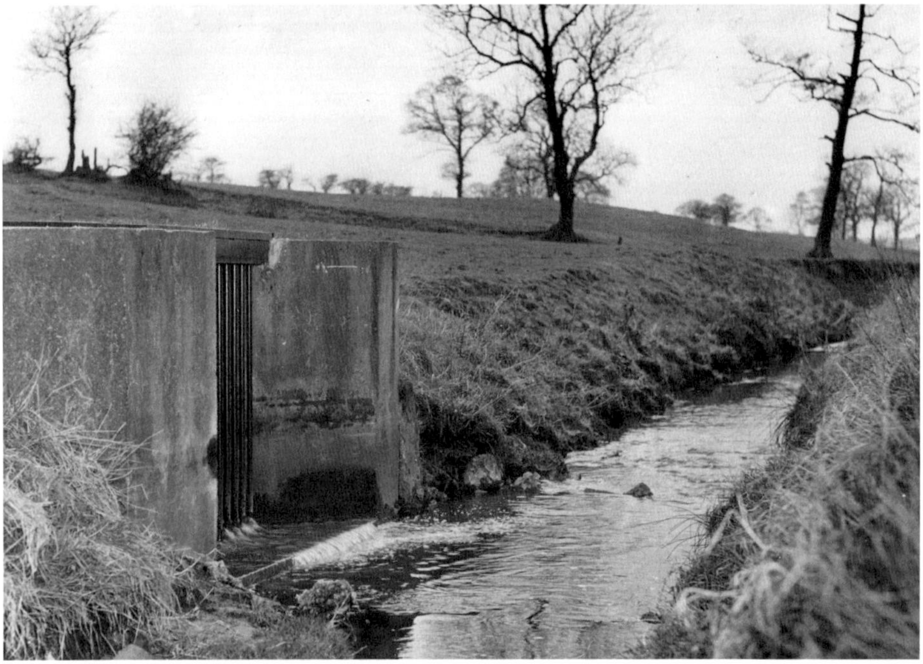

Water that had been through the company's interceptor was discharged into the North or Leighton Brook. Very few employees ever ventured this far.

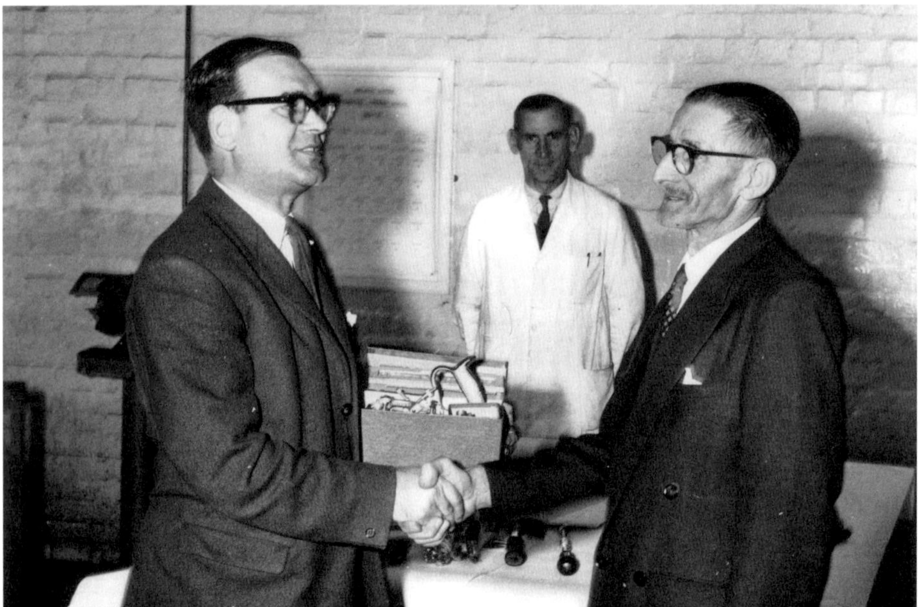

Ronnie Dyson presenting Ernie Ward, the first apprentices' instructor, with a retirement gift. Arthur Hammersley is looking on. The much respected Dyson was one of the first seven men sent from Derby to establish Rolls-Royce in Crewe in July 1938.

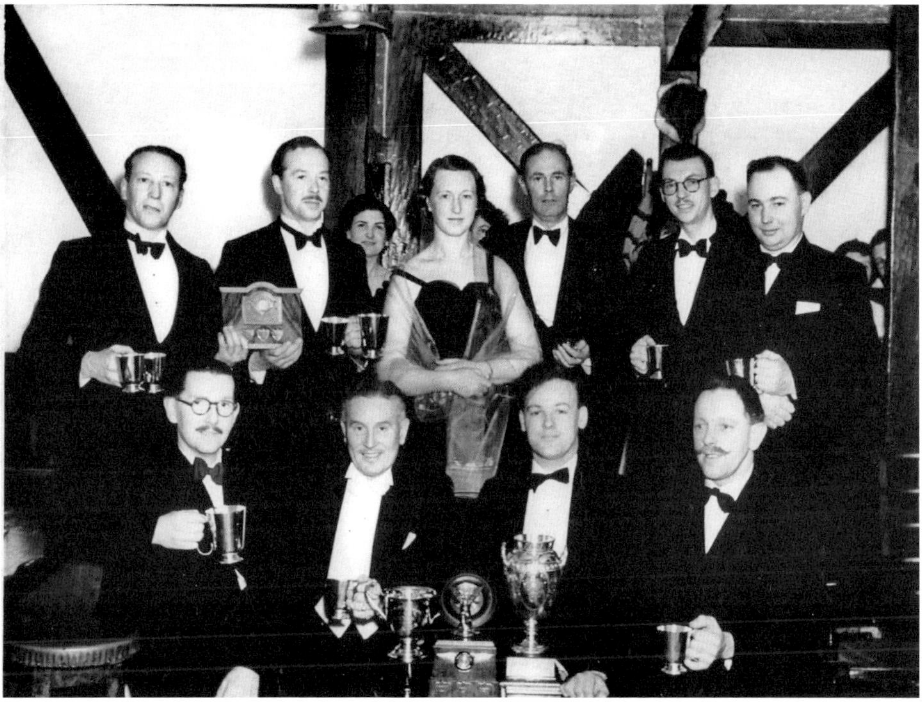

Rolls-Royce (Crewe) Motor Club 1952. Back row; N. Gawthrop, R. Jackson, Mrs Grylls, H. Grylls, R. Spencer, W. Sexton: Front row; K. Blagg, D. Kershaw, B. Leverton, C. Dix.

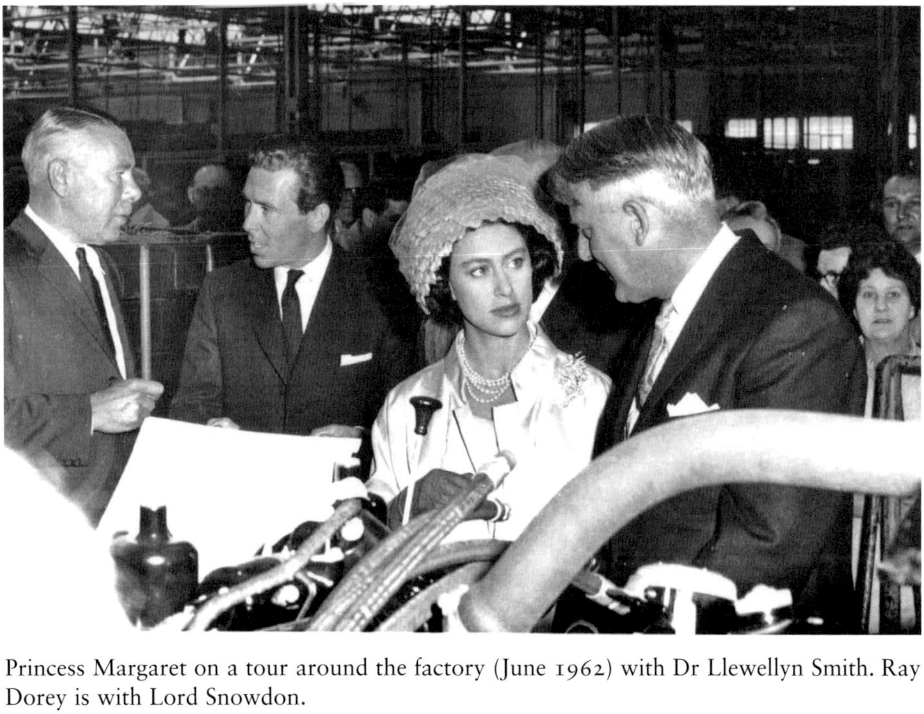

Princess Margaret on a tour around the factory (June 1962) with Dr Llewellyn Smith. Ray Dorey is with Lord Snowdon.

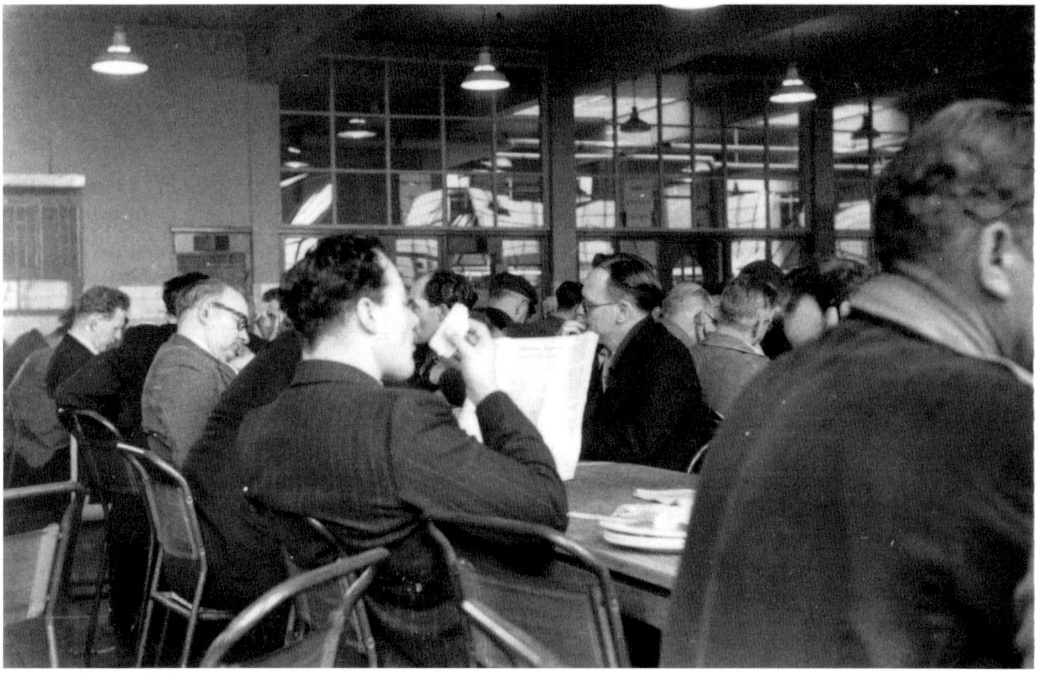

The staff canteen in the fifties. There were at least four different dining locations for various grades of employees.

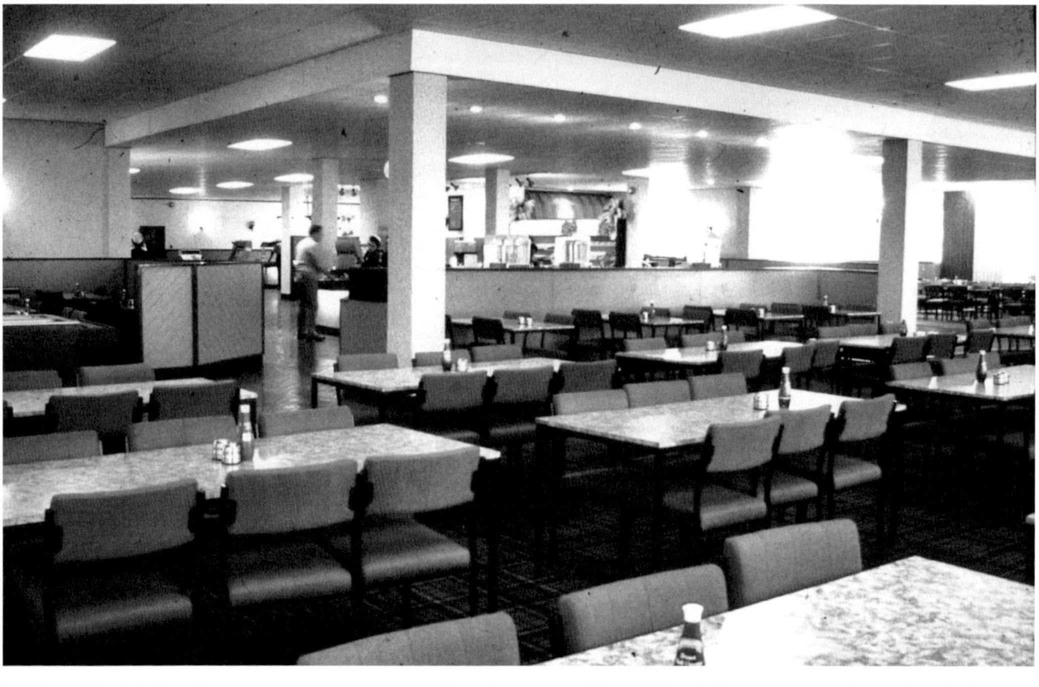

In the mid-eighties the canteen was renovated and renamed Brooklands Restaurant where (nearly) all dined together.

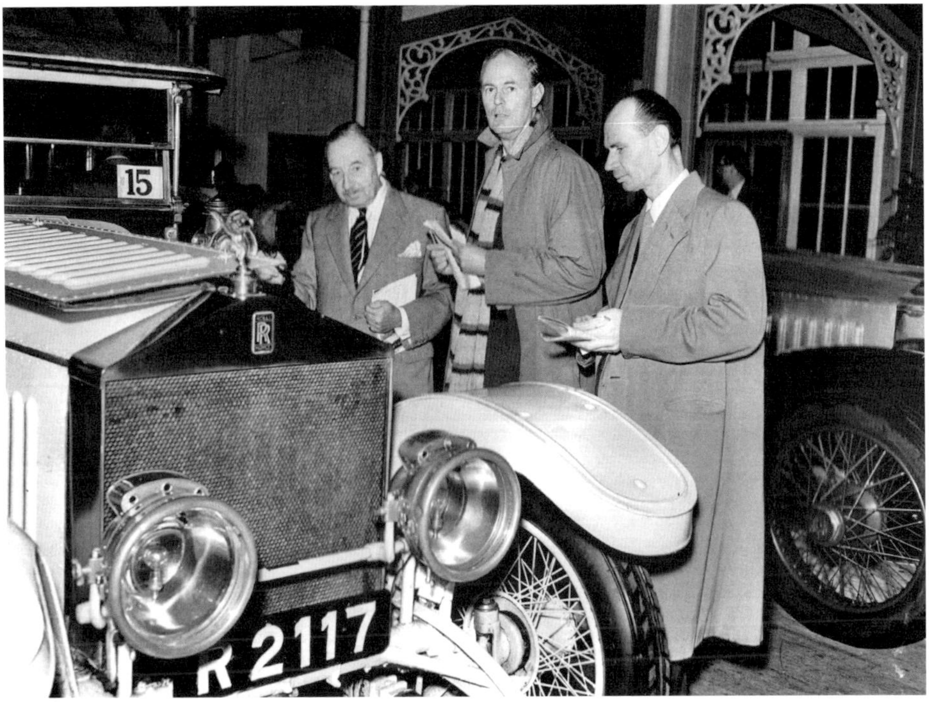

Harry Grylls, chief engineer and John Blatchley chief stylist.

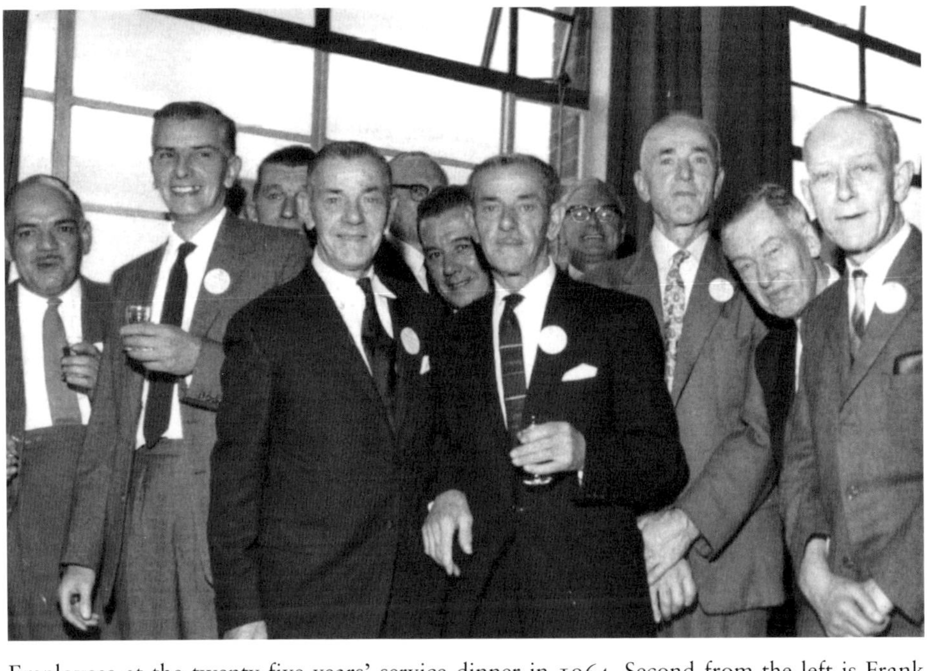

Employees at the twenty-five years' service dinner in 1964. Second from the left is Frank Clark who helped to install Merlin variants into aircraft at wartime RAF stations. The twins in the centre are the author's father and uncle. Another uncle is at the rear.

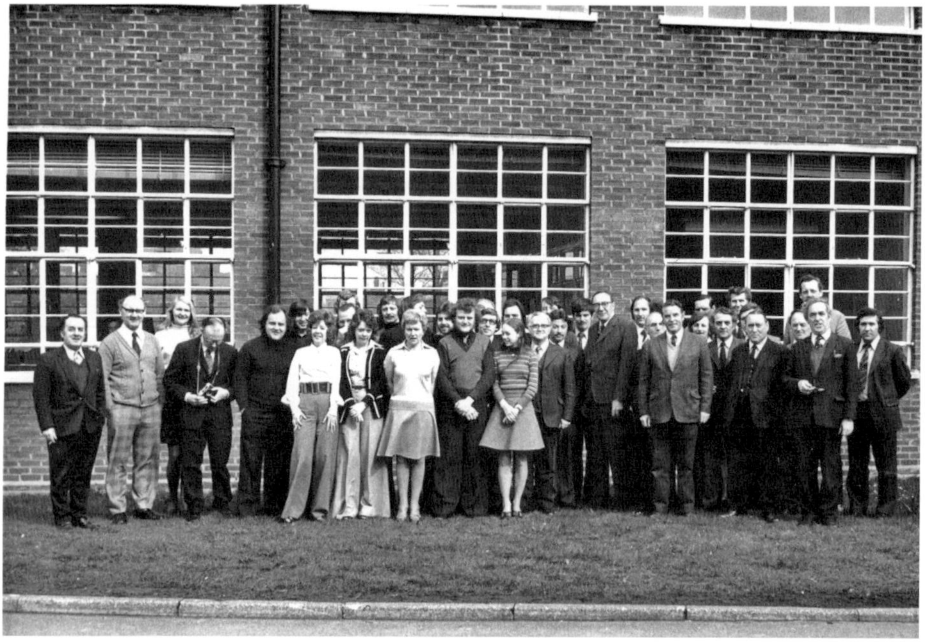

Draughtsmen outside of the detail drawing office. Towards the centre wearing spectacles is Harold Peak, chief draughtsman and a scratch golfer.

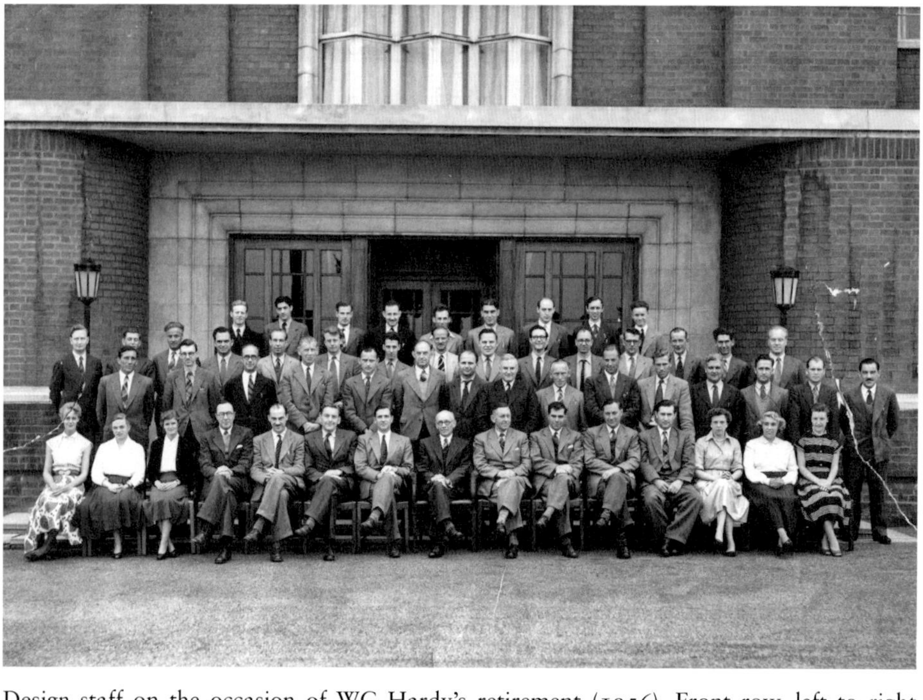

Design staff on the occasion of WG Hardy's retirement (1956). Front row, left to right are: A. Earnshaw, M. Forster, S. Williams, H. Peak, G. King, F. Tarleton, V. Stanbury, Hdy, R. Davis, J. Phillips and H. Jeal, E. Leaver, M. Bonsall, C. Jenner, M. Jones.

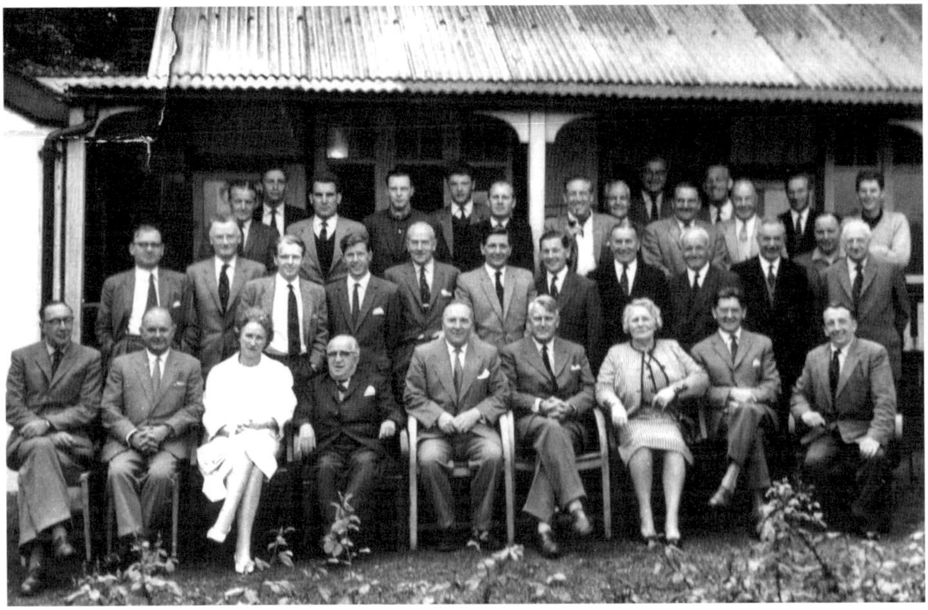

Crewe RR golf section. Centre of the front row are A. Brogden and Dr Llewellyn Smith. There was a rumour that joining the golf section was the best avenue for promotion.

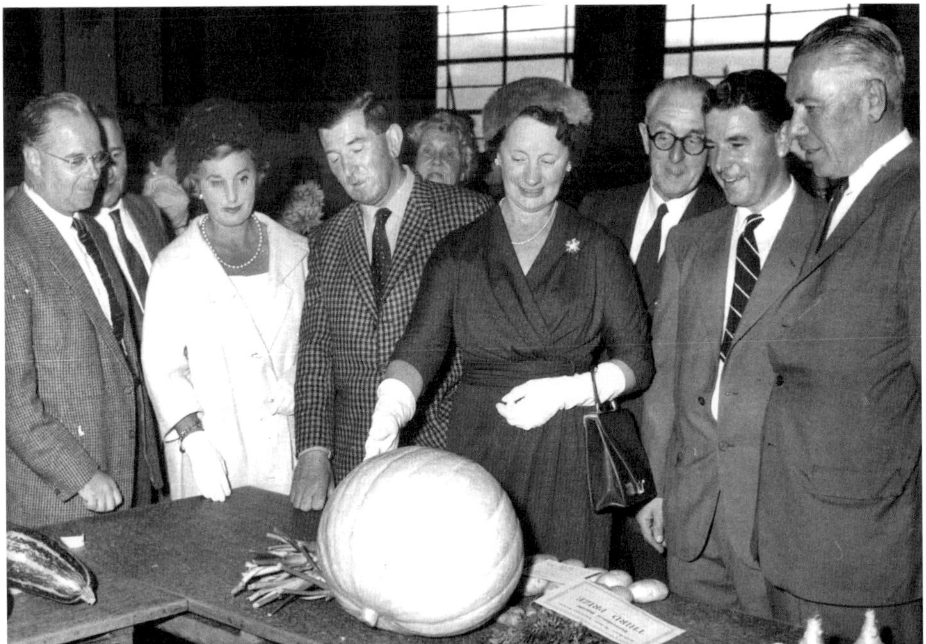

RR Horticultural Show (*c.* 1958) Left to right: W. Elliott, Joan Elliott, Jack Hewitson, onetime mechanic to Sammy Davis the Bentley racing driver, Mrs Dorey, J. Valentine, B. Jones and Ray Dorey. This latter man was responsible for installing the RR Merlin engine into the Mustang.

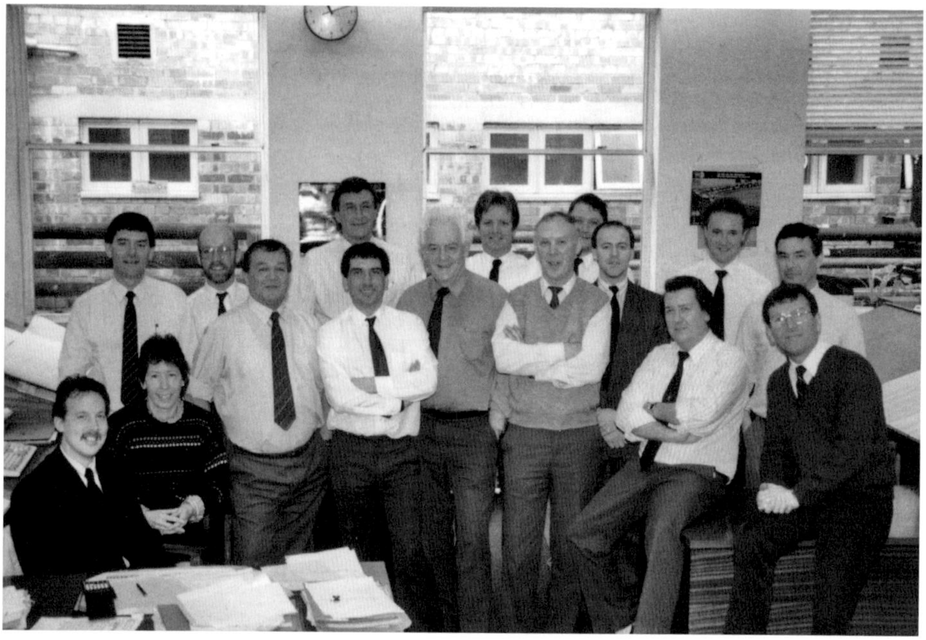

Estimating department in the seventies. Included are K. Birch, P. Dyer, J. Walley, S. Tomlinson and R. Hodgkinson.

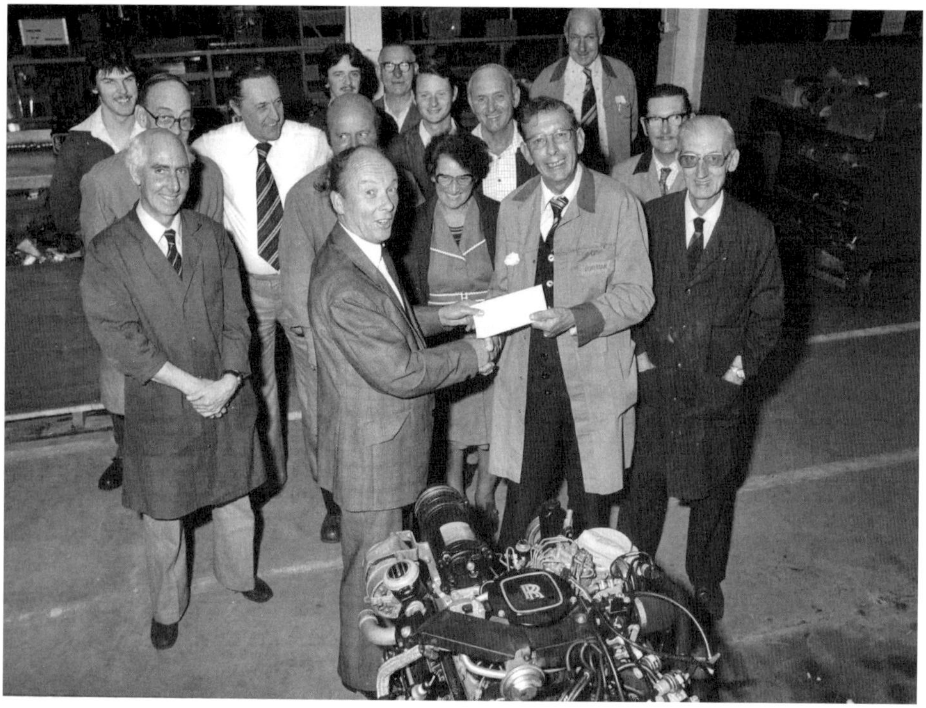

John Fox (experimental) presents S. Taylor with a retirement gift. Others in the group are E. Pointon, C. Shaw, and Bill Elliott.

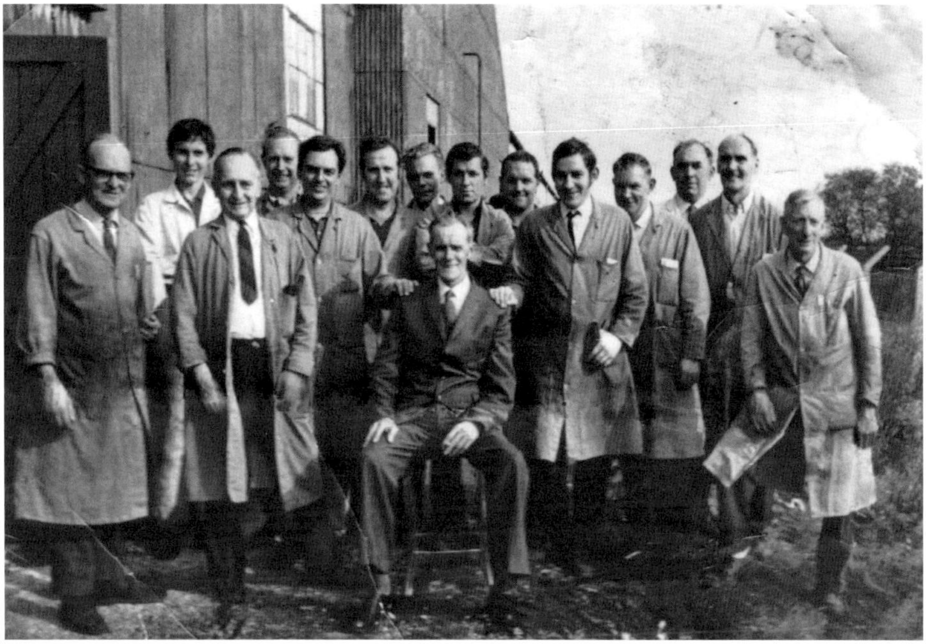

Machine tool maintenance (machining) outside of their workshop adjacent to the Chester line (*c.* 1970). Bob Harvey and Rueben Gibbons extreme left and right respectively were pre-war employees who migrated into Crewe.

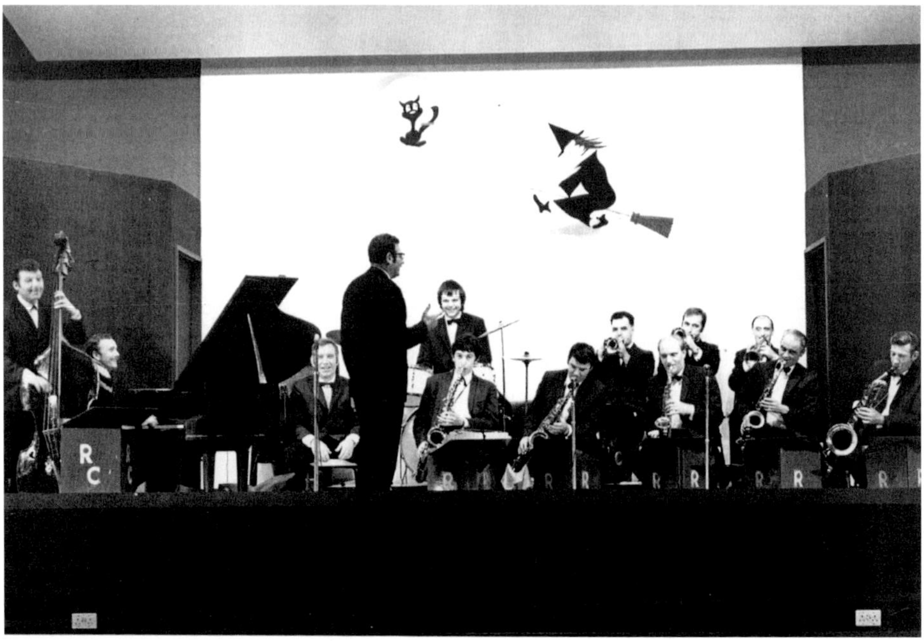

Ralph Cowdall's Band in the Silver Starlight Room 1972. Ralph was a toolroom inspector at Pym's Lane. Other employees were M. Bourne (piano), A. McKenzie and R. Latham (tenor sax), J. Jones (trumpet), A. Vickers (trumpet) and E. Hopley (baritone sax).

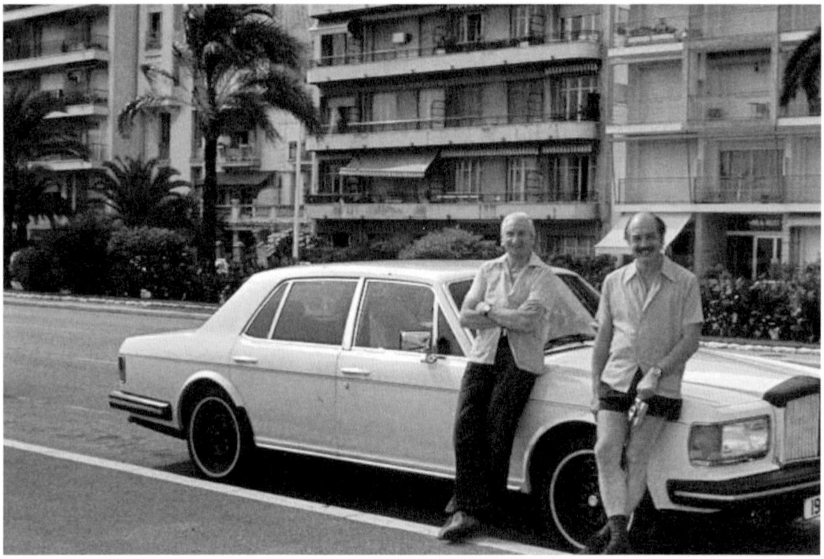

Development engineer Derek Coulson (in shorts) taking a break while on the way to Nice for the launch of the Silver Spirit in September 1980.

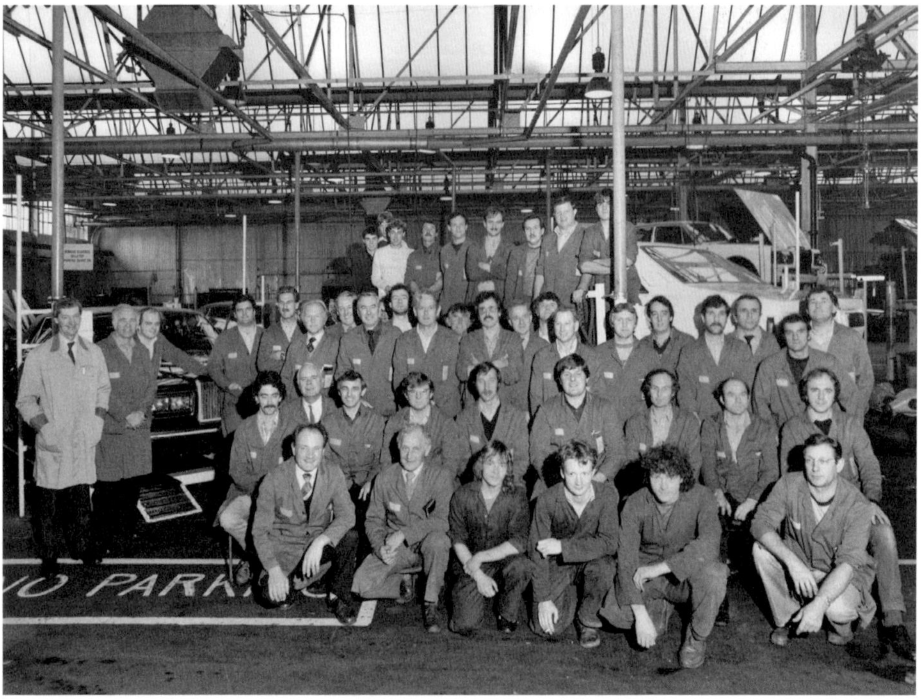

Road test in the mid-eighties. A Silver Spirit destined for the USA is left of the assembled group. B. Dickinson, foreman is on the extreme left. G. Hulse, (final test) is on the left of the front row with J Bolderstone (final test) immediately behind while J. Taylor (future foreman) is behind him.

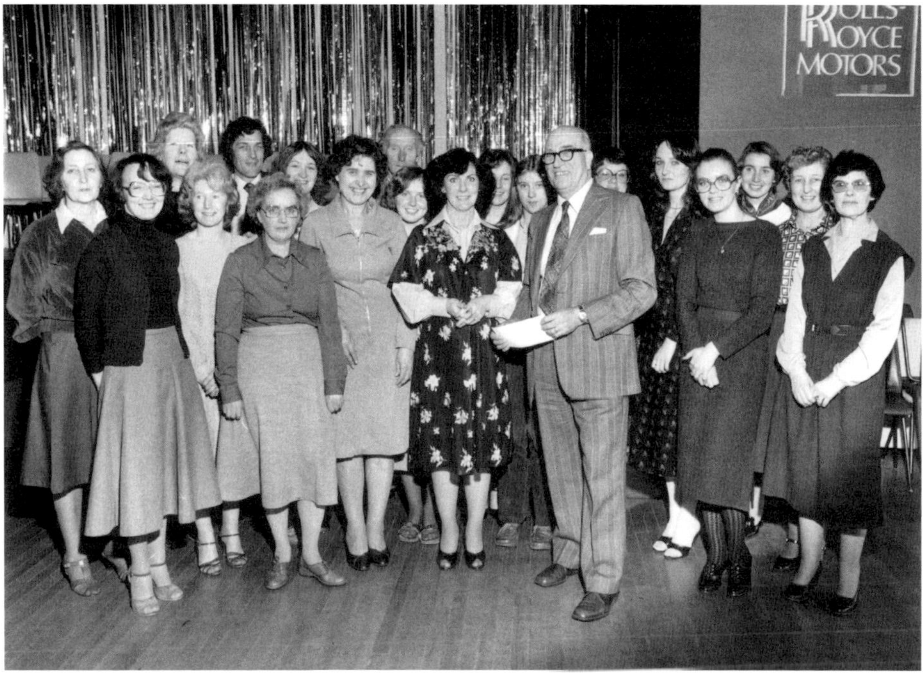

Tim Friesner, a senior planner and the ladies of the planning dept. Among the group are P. Lovell (receiving present), D. Dyer, V. Wardell, P. Taylor, D. Coates and L. Beckett.

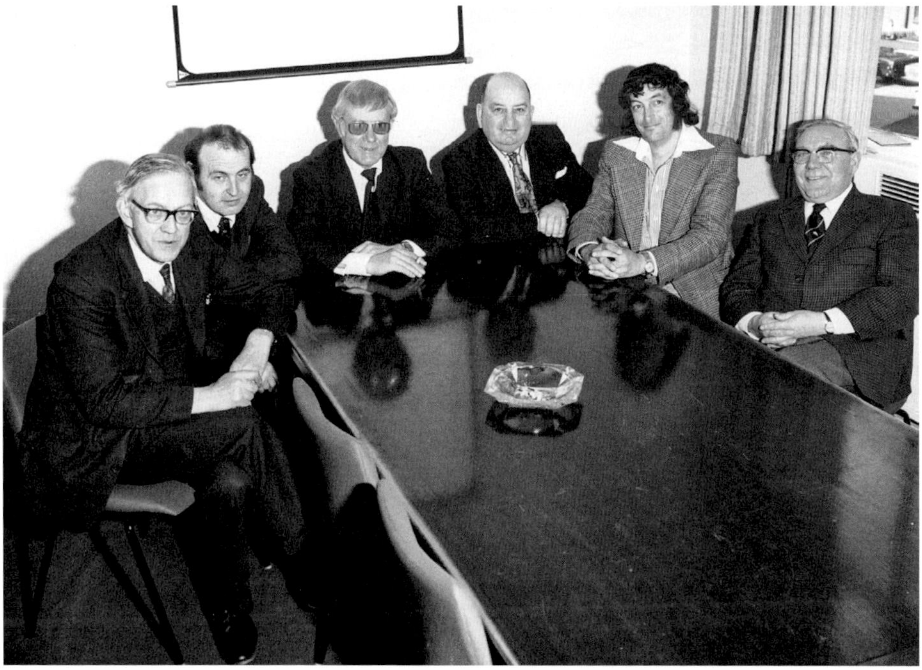

D. Long (planner), A. Flood (quality), G. Fovargue (spares), C. Tew (material control), K. Simcock (jig and tool) and E. Dodd (progress), all senior staff union reps.

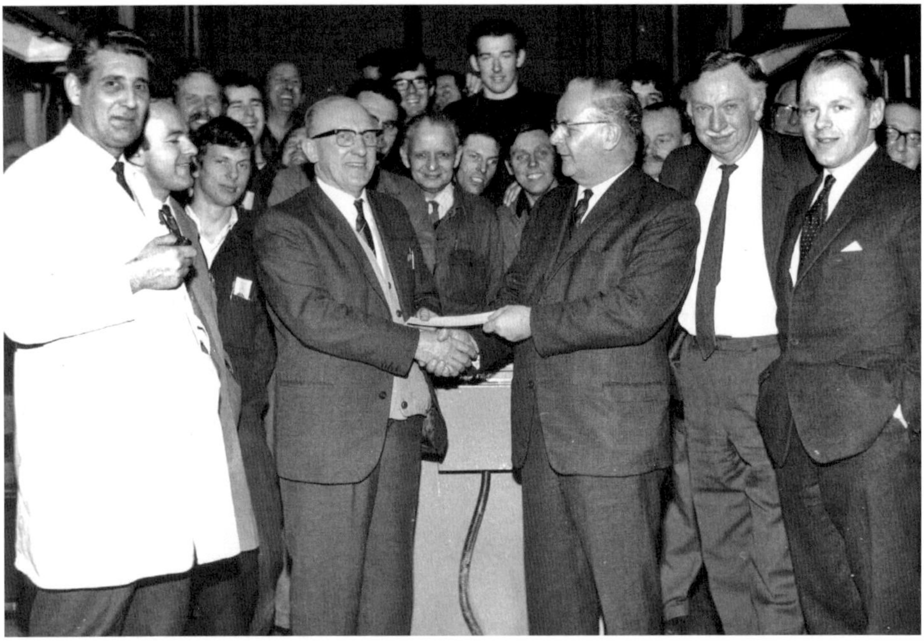

George Eardley (electrician) the only holder of the VC at Pym's Lane. On the right are Syd Torr and Tom Neville (finance director).

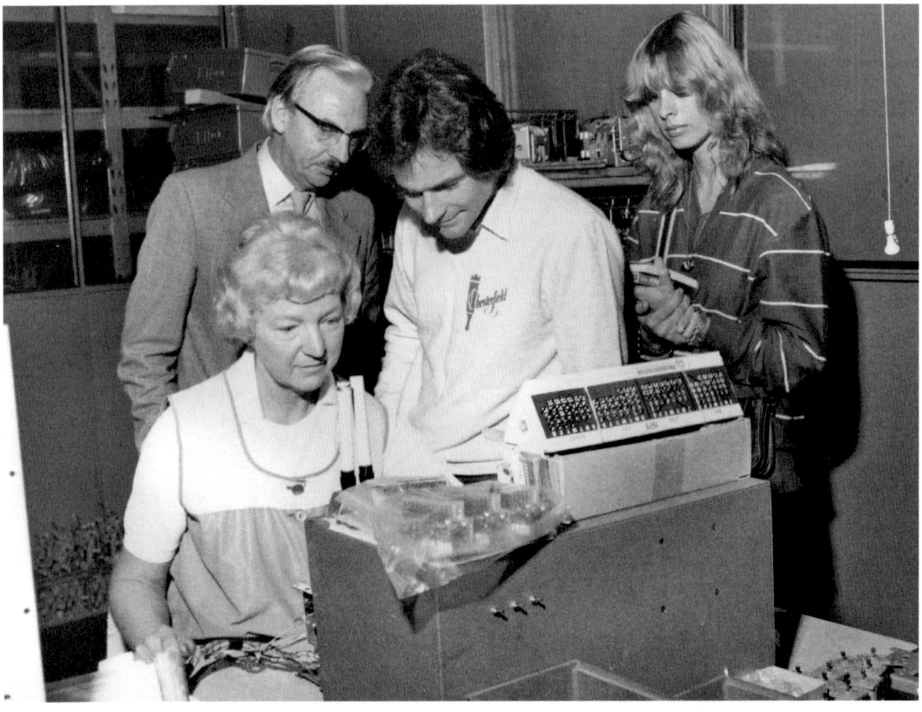

Barrie Sheene, world champion motorcycle rider visiting the factory prior to purchasing a RR Silver Shadow II. Marjorie Goodwin is testing an automatic a/c unit with Reg Spencer (for many years a designer before becoming chief inspector) looking on.

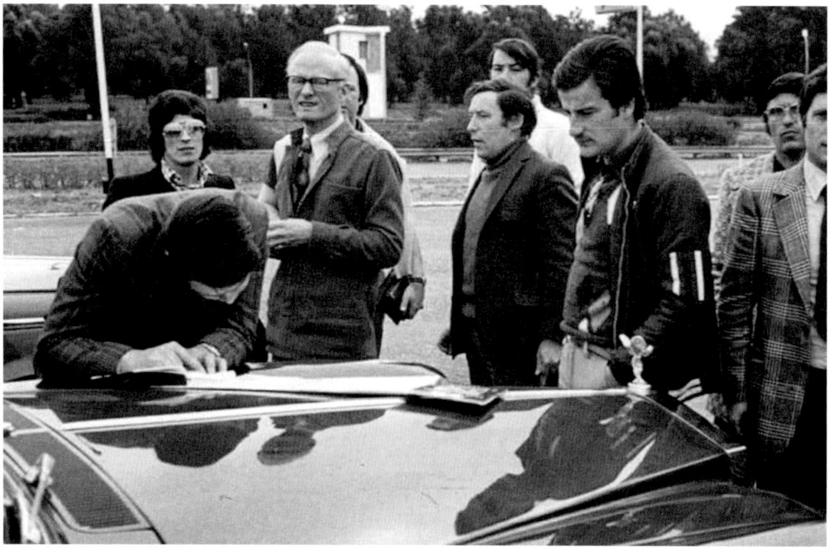

11 January 1975 near to Monte Casino on the way to the Camargue launch. Bruce Sutherland studies a map with Roger Cra'ster (sales) and Jim Arrowsmith (service manager) along side.

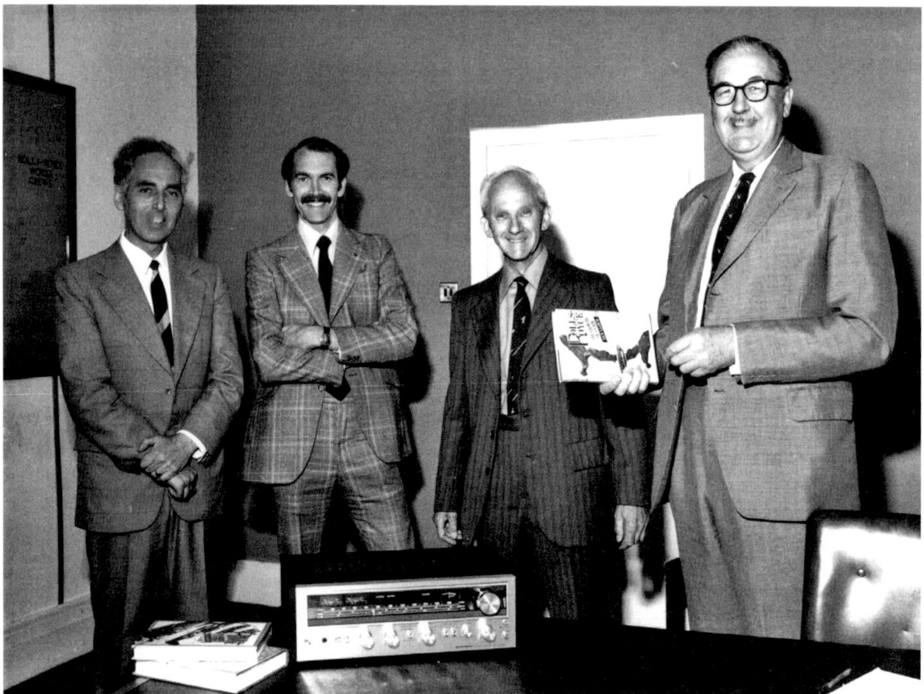

Frank Culley's retirement in 1979. Frank was one of the first employees engaged by Crewe RR in August 1938. From left to right: Mike Parsons, Barry Treadgold, Frank Culley, John Hollings (chief engineer).

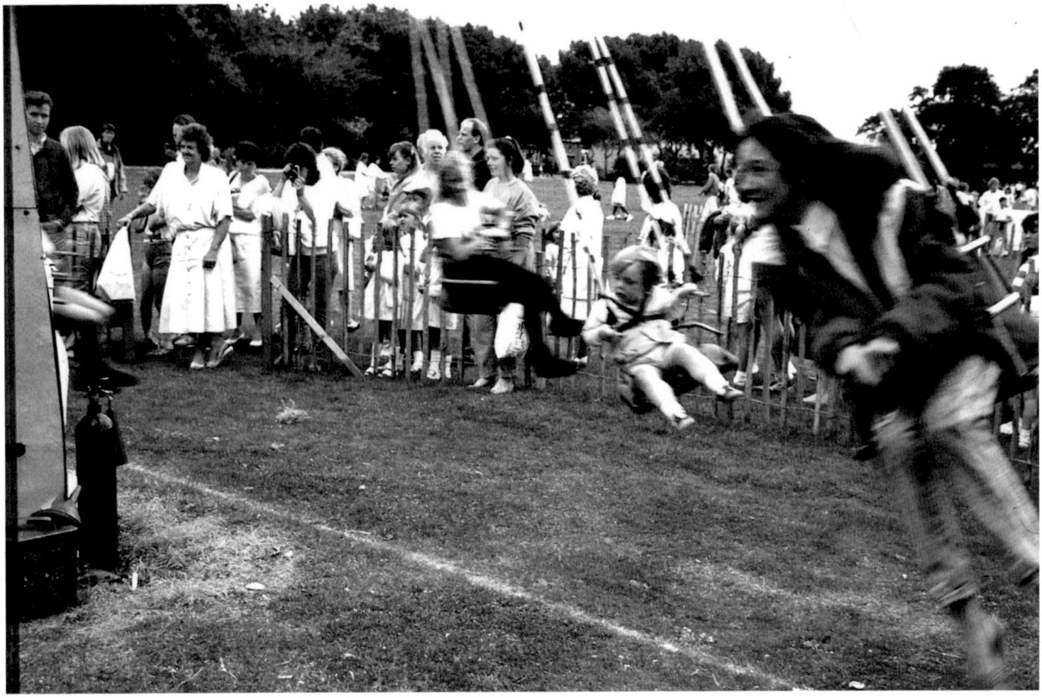

Chair o' planes was a perennial attraction at the children's field day.

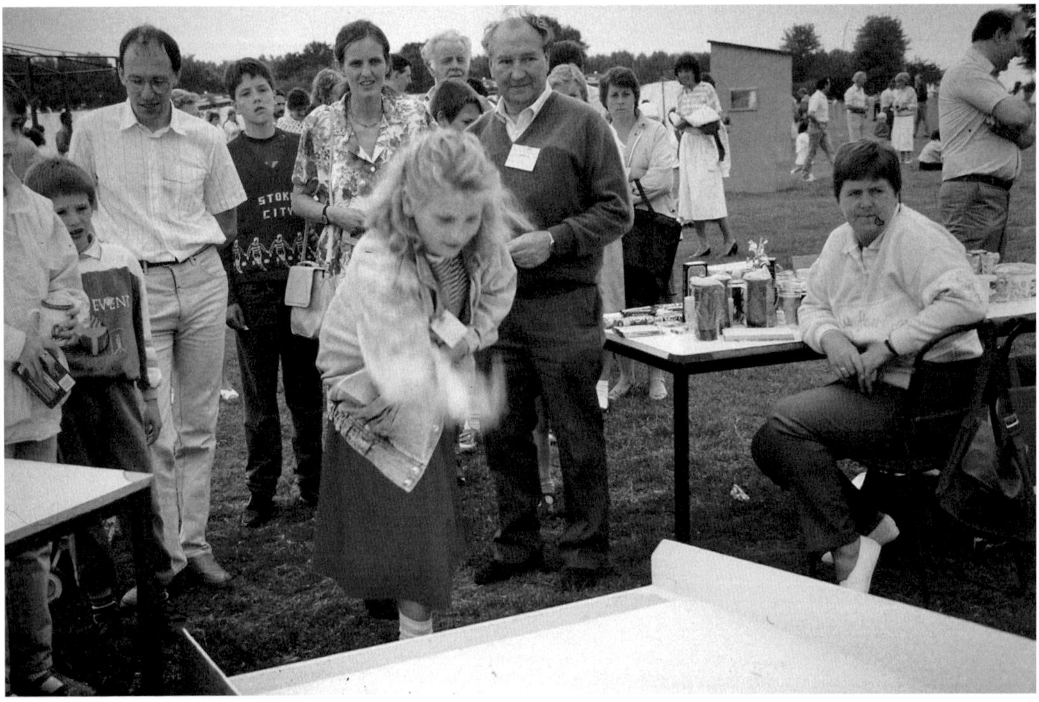

Norman Evanson, of 153 department, a volunteer helper (*c.* 1980).

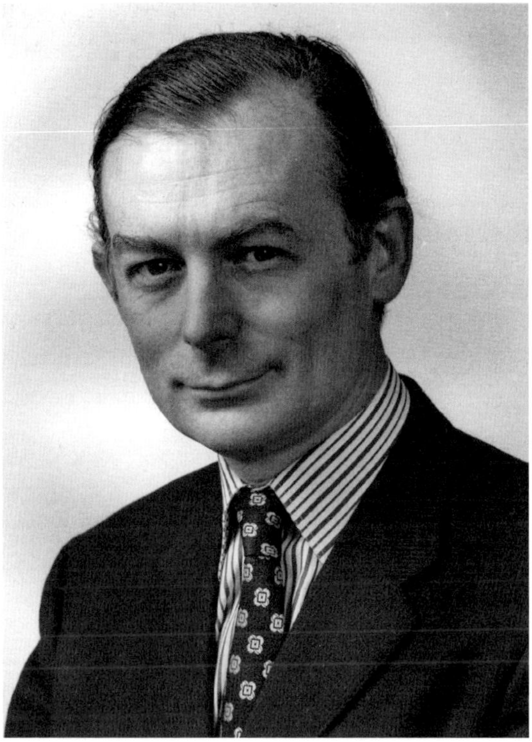

David Plastow, Group Managing Director who successfully negotiated the flotation of Rolls-Royce Motors in the early seventies.

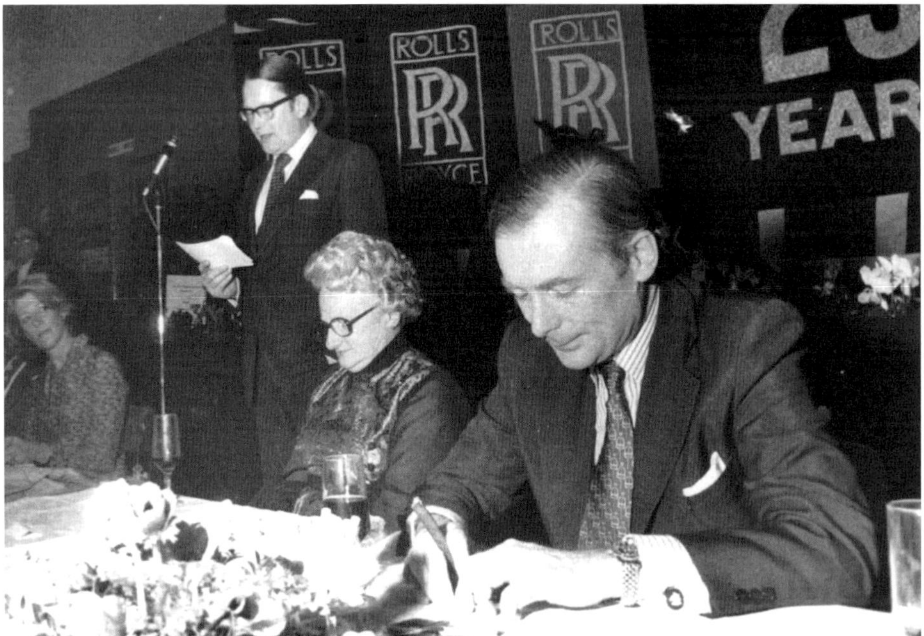

George Fenn (MD) rousing the troops at a twenty-five year dinner. Listening is D. Plastow, knighted in 1986.

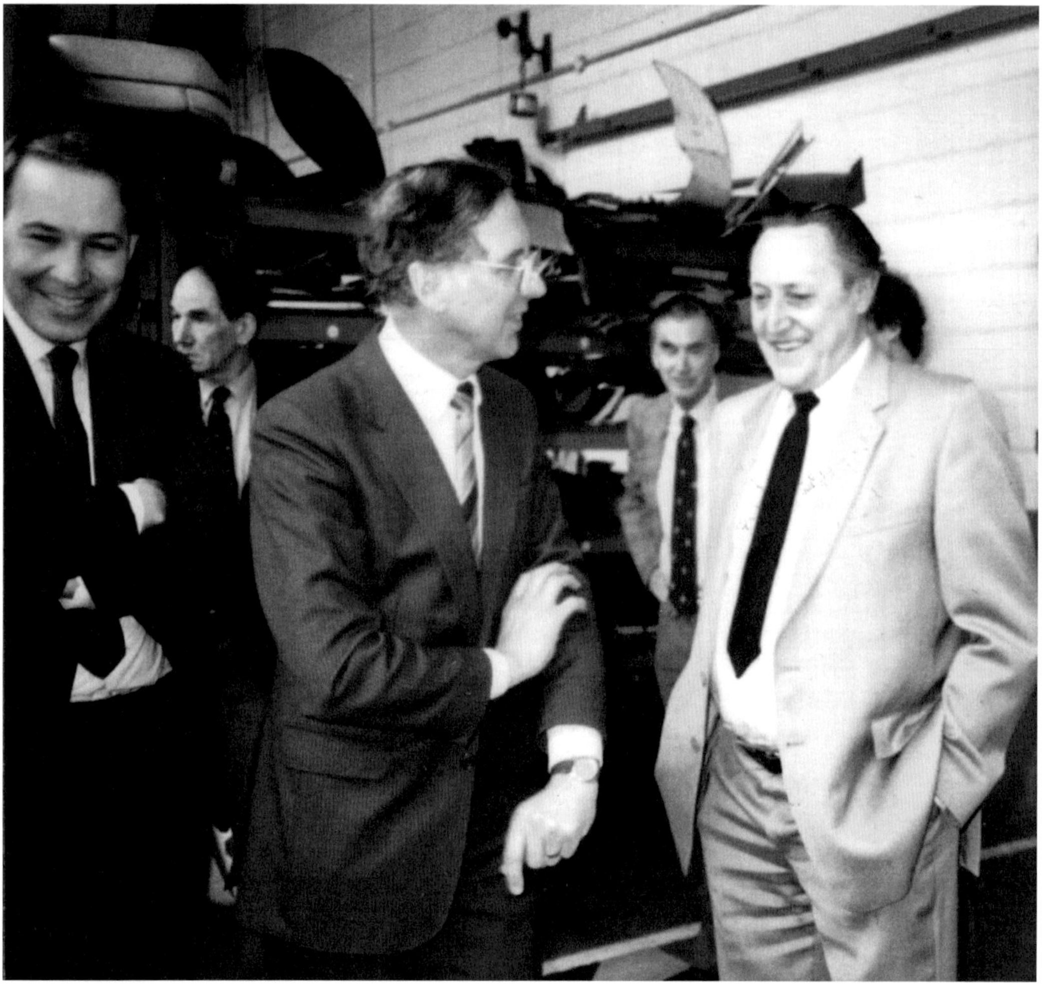

Nick Colbourne, Jock Knight, Geoff Moreton, MacFisher and Ern Pointon in the engineering block (*c.* 1982). Geoff and Ern started to work at Pym's Lane on the same day in 1949 as centre lathe turners.

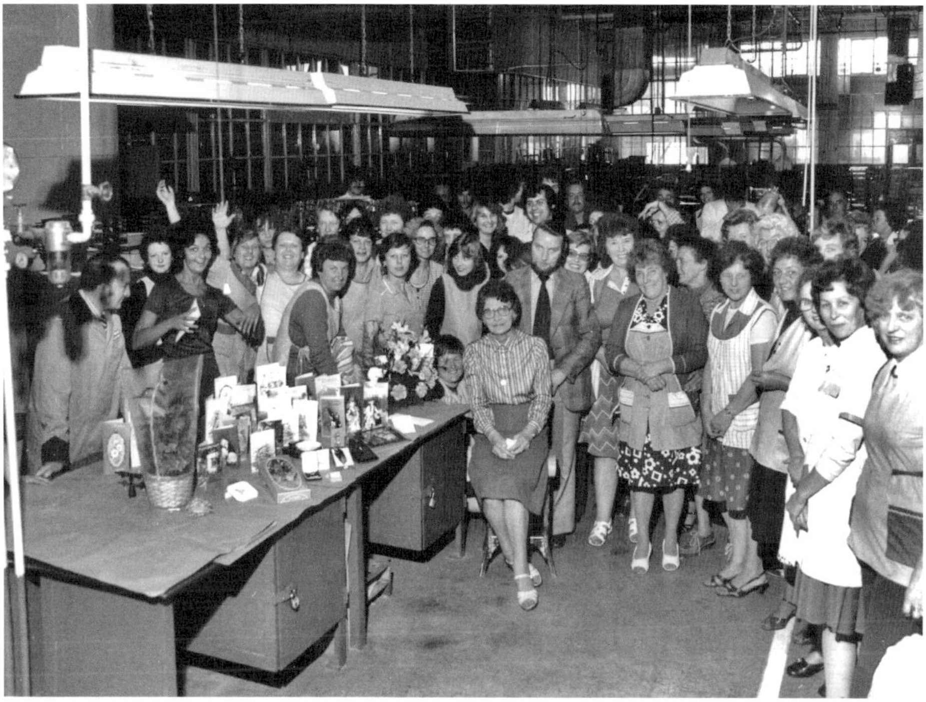

Ladies of the investment foundry wax room in the late eighties.

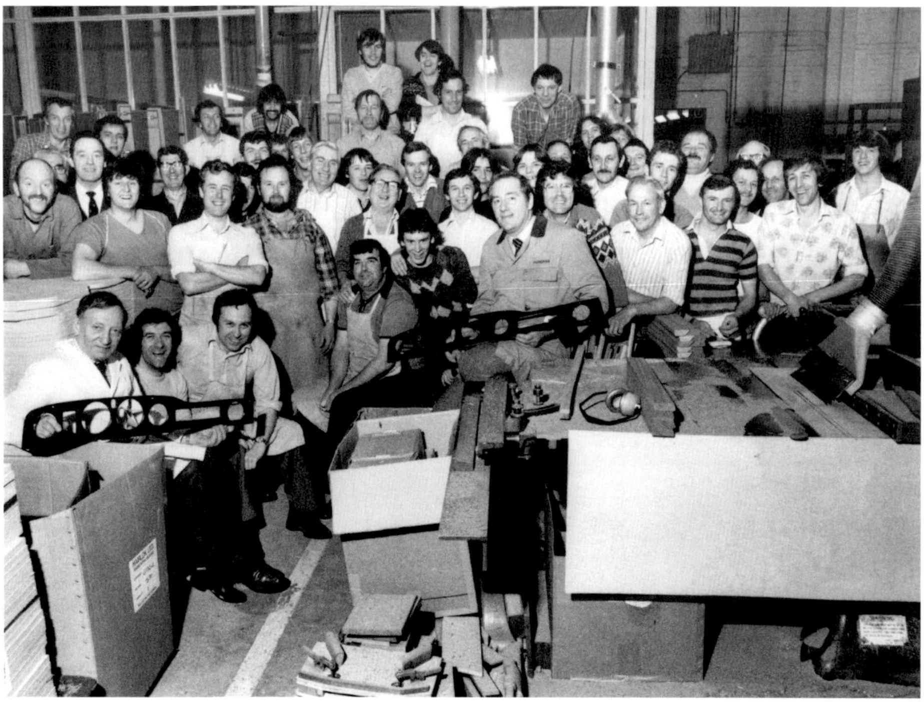

Tony Clayson (foreman) in the midst of workers from the wood shop.

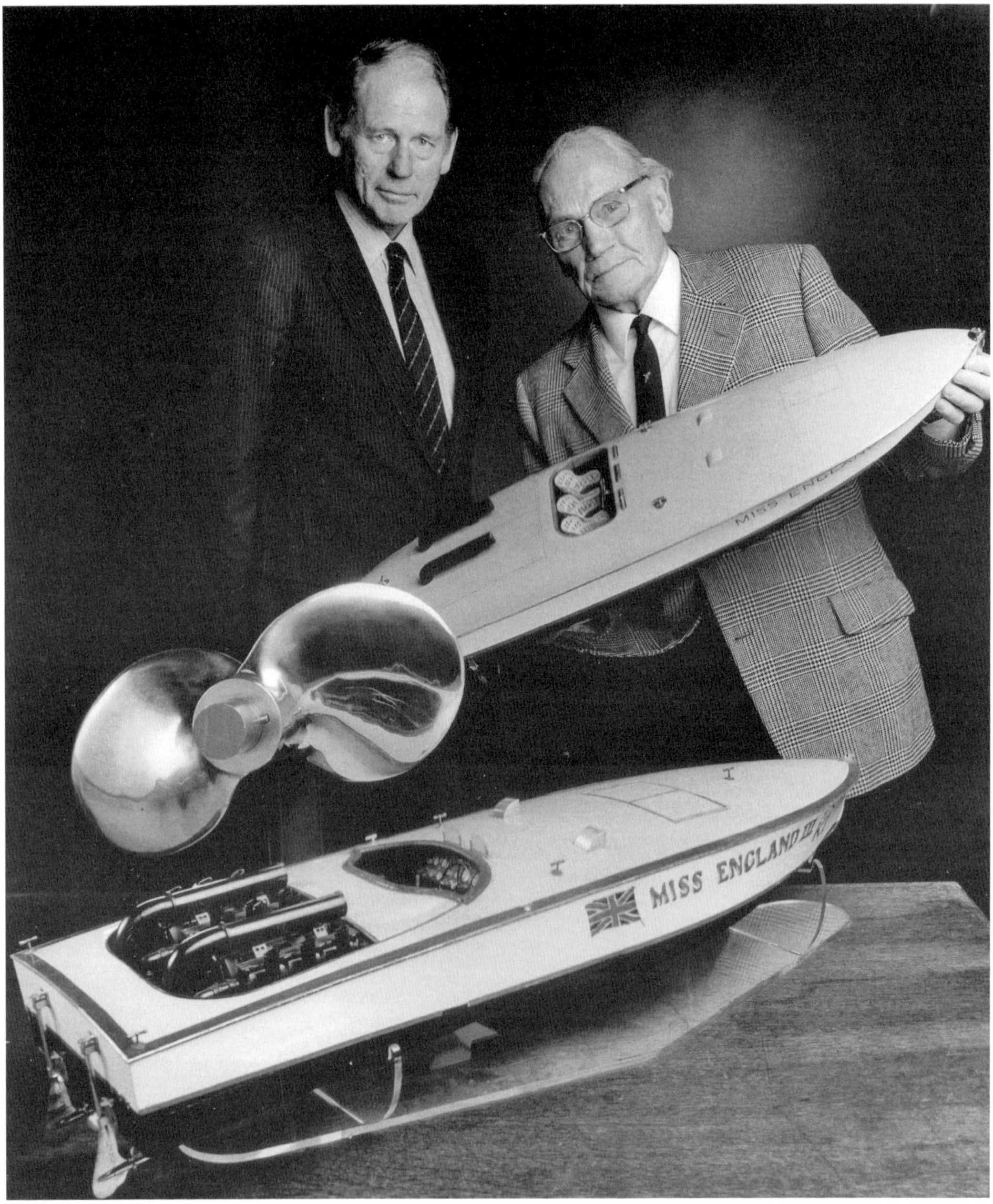

Dick Perry (chief executive) and Dick Garner (*c.* 1983). The latter man was one of the original Derby seven from July 1938. He had also been mechanic for Kaye Don on the Miss England speed boat project, hence the model. Garner efficiently carried out many duties at Pym's Lane including chief engine tester, materials director and chairman of Mulliner Park Ward. He retired in 1974.

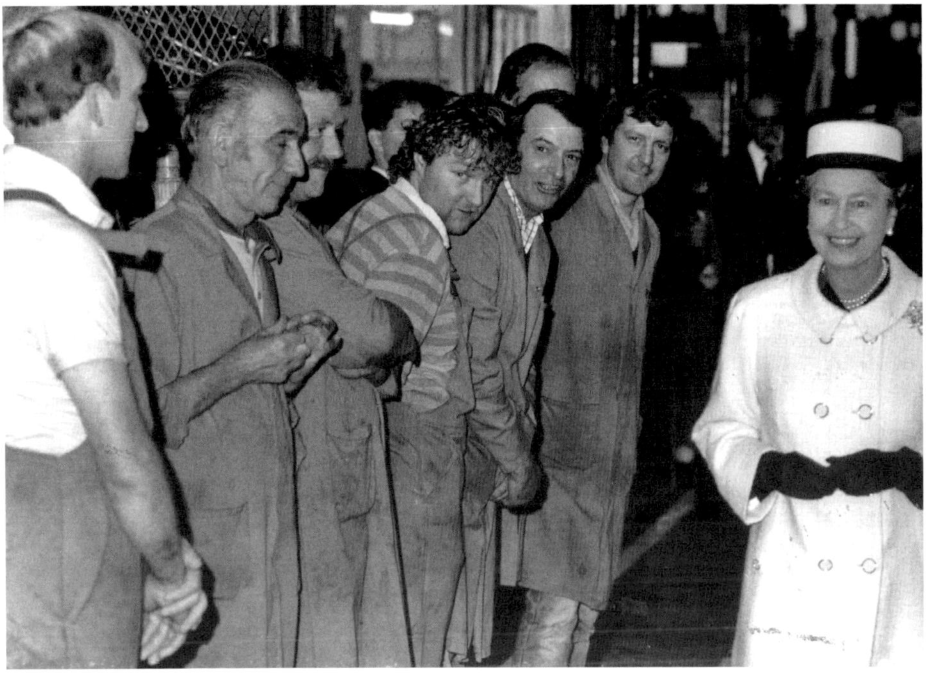

In July 1987 Pym's Lane factory was on the itinerary of the Queen when she visited Crewe to celebrate the 150th anniversary of the arrival of the railway.

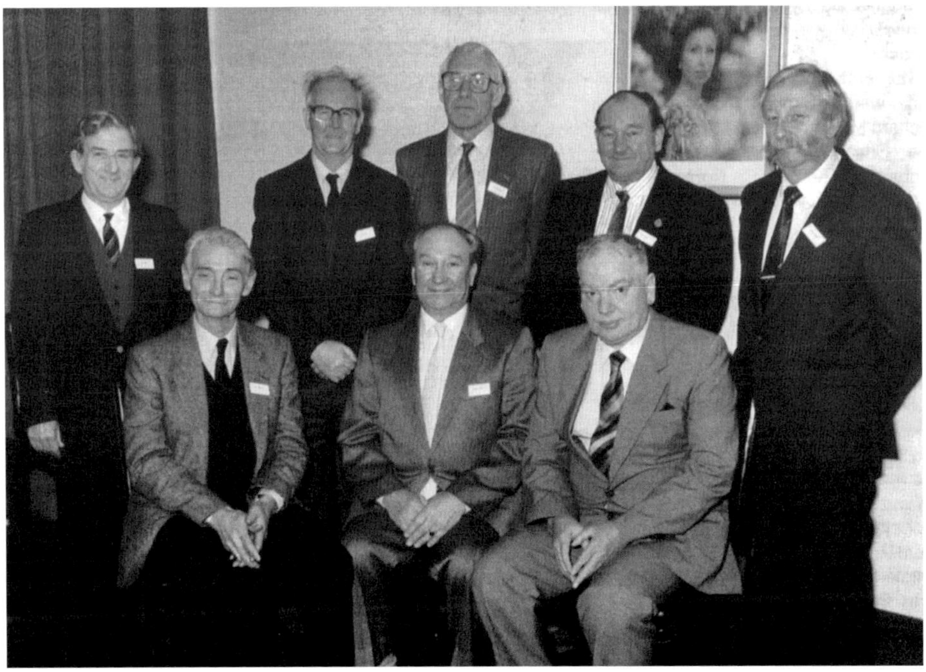

Celebrating fifty years service in 1991 are J. Knapper, N. Evanson, A. Sanders, T. Whittaker, R. Reade, J. Morris, T. Boffey and E. Cornes.

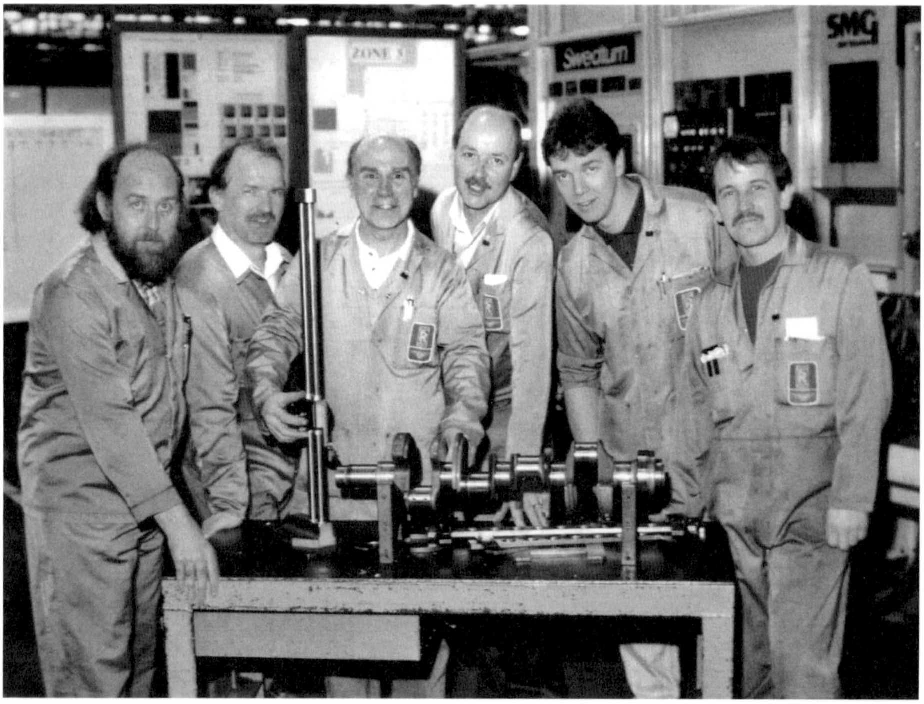

John Rhodes with one of the teams formed to initiate the *Strive for Perfection* project in 1993.

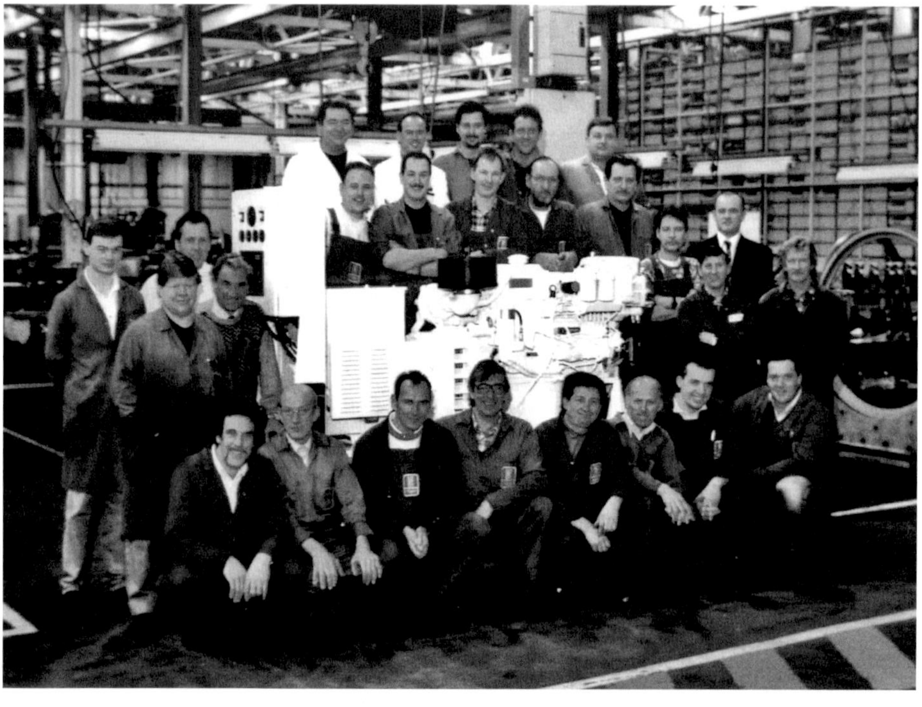

The special engine division with an FD 12 power unit (*c.* 1993).

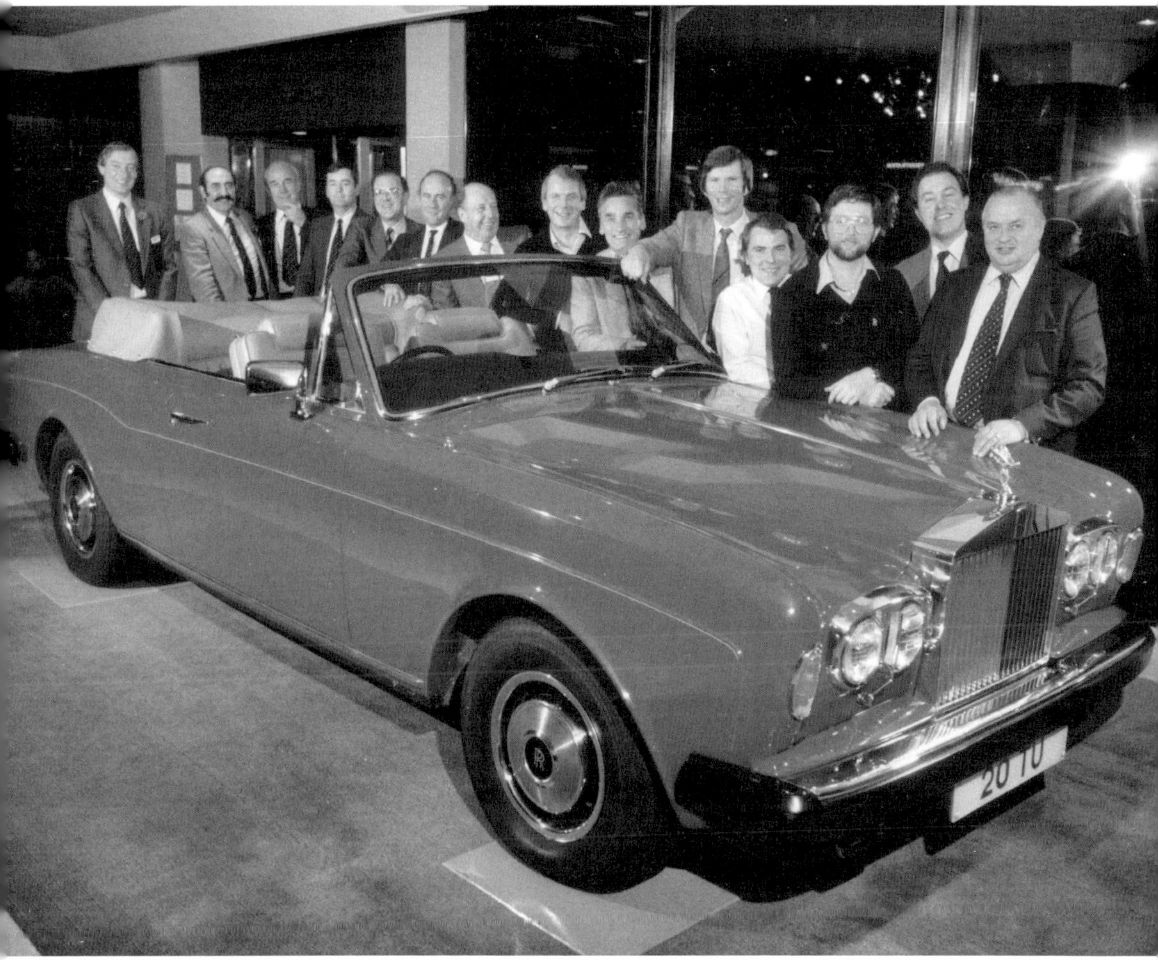

Bernard Preston service director alongside a RR Corniche with (among others) A. Read, G. Fry, J. Cowell, and S. Hawley, all service engineers (*c.* 1984).

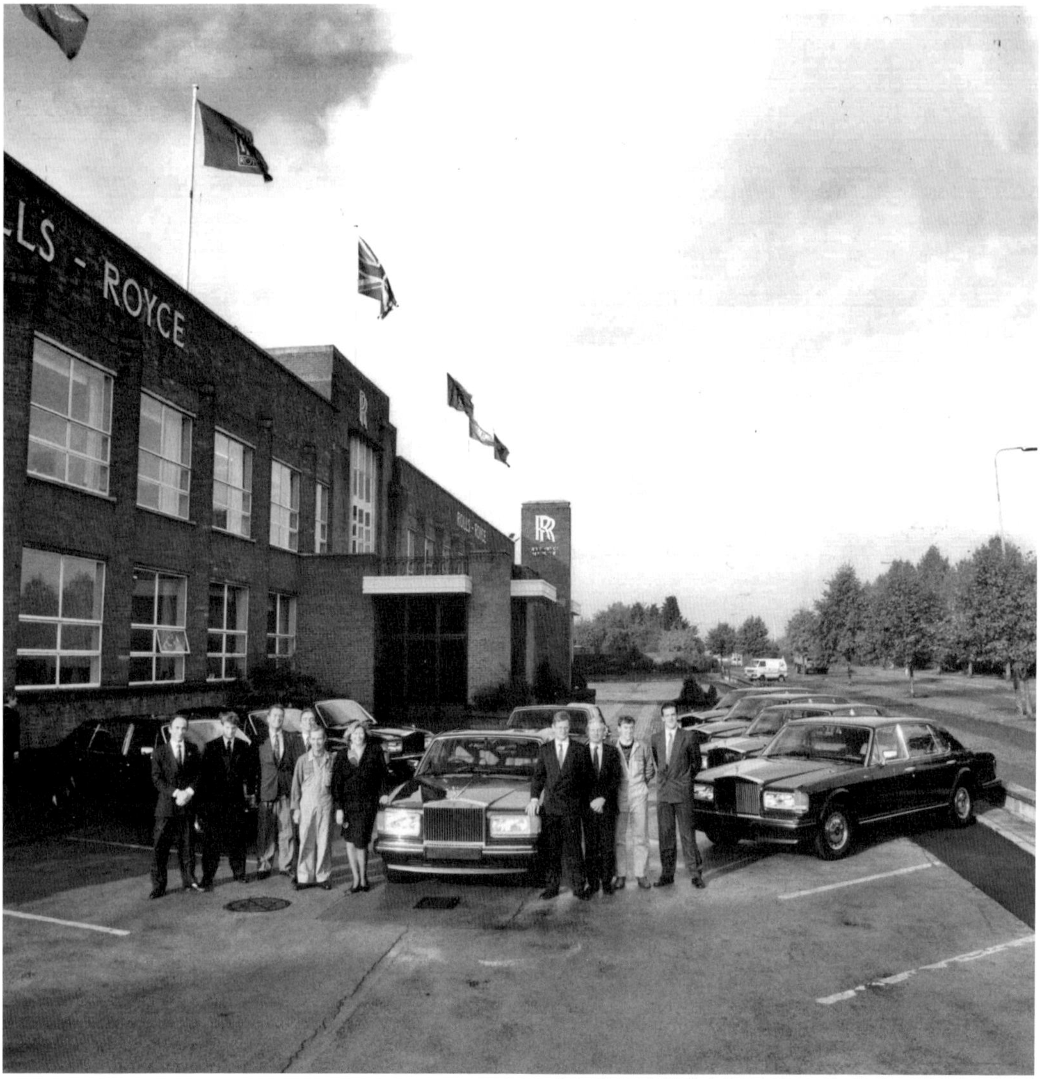

Looking west along Pym's Lane with P. Cartwright, P. Hardstaff, G. Dodgson, J. Brodie, W. Slater and Sue Langer around a RR Silver Spirit.

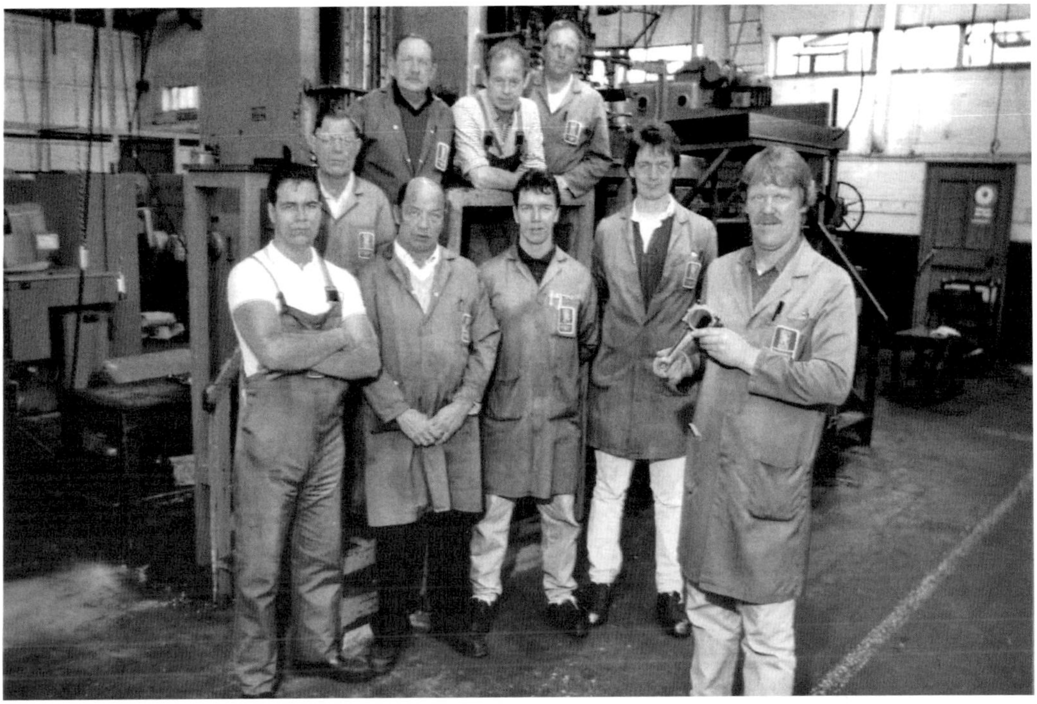

A group of workers from the con-rod bay prior to the demise of engine manufacturing.

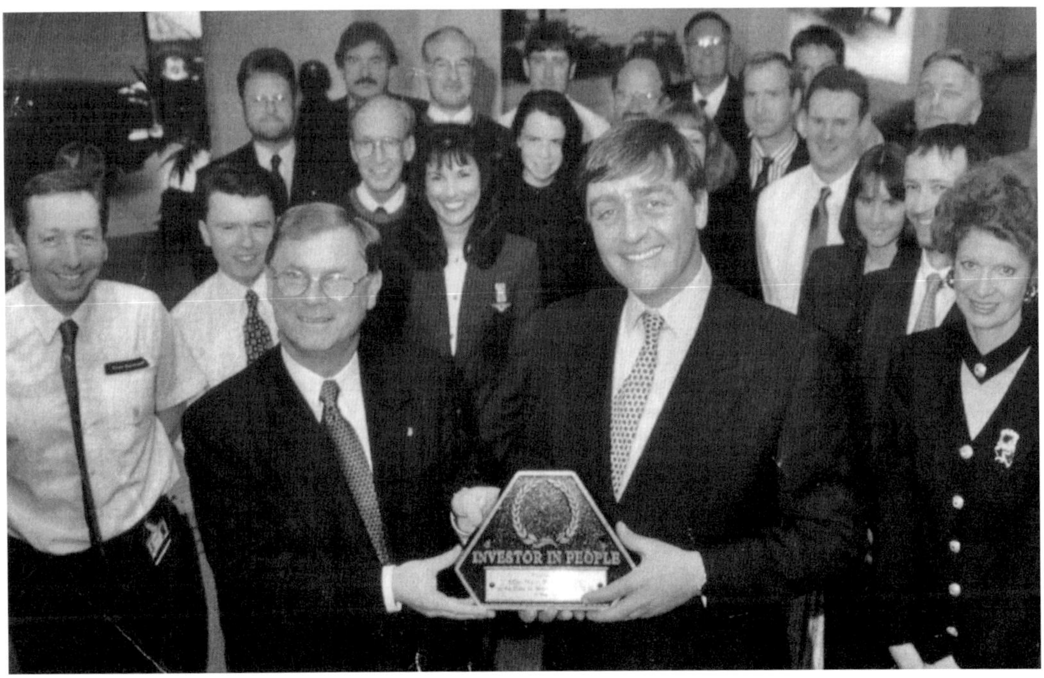

The Duke of Westminster presenting Graham Morris (chief executive) with the *Investors in People* award. Sue Evans and D. Davenport are among the onlookers.

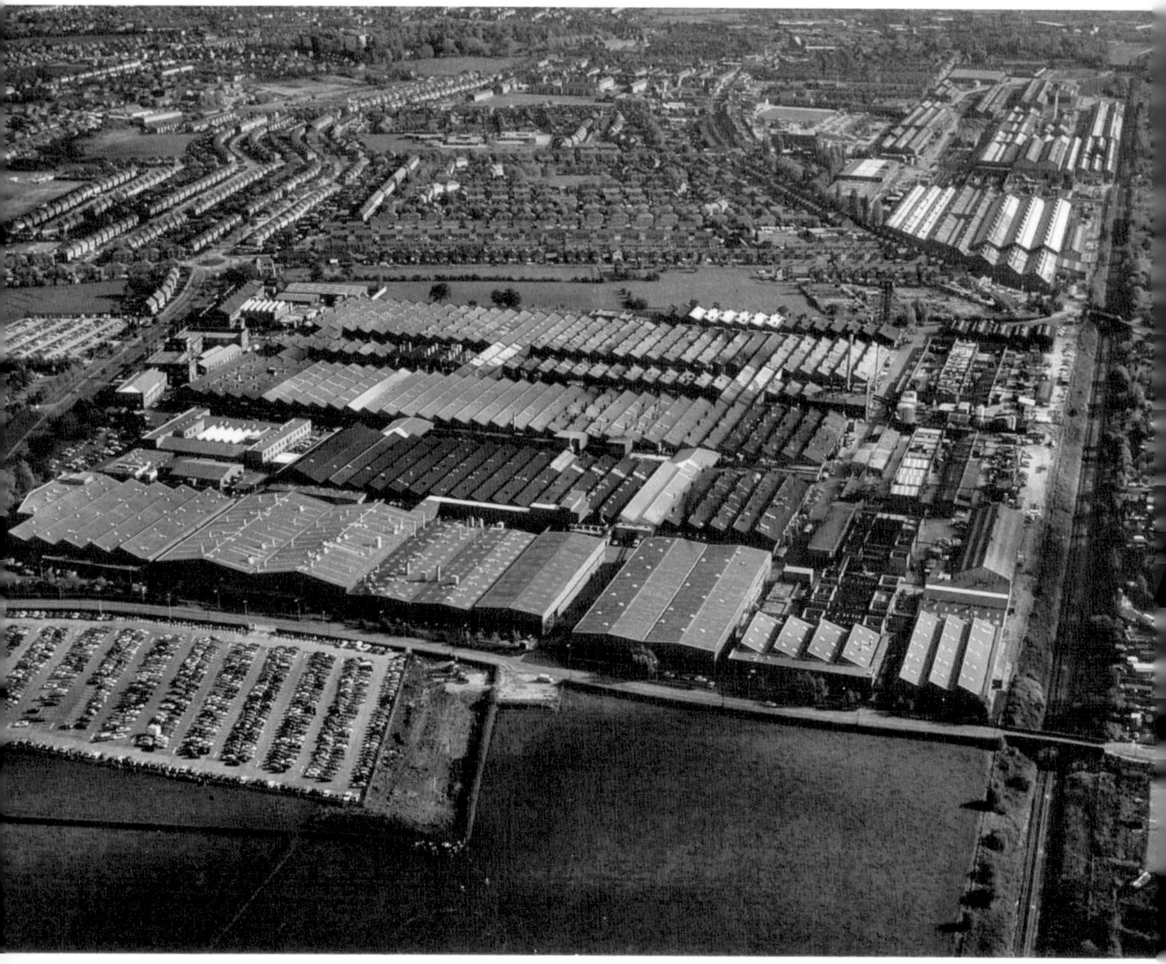

Aerial view of the factory looking east in the late nineties. An extra car park (sign of the times) was built on the west of Sunnybank Road in the late seventies.

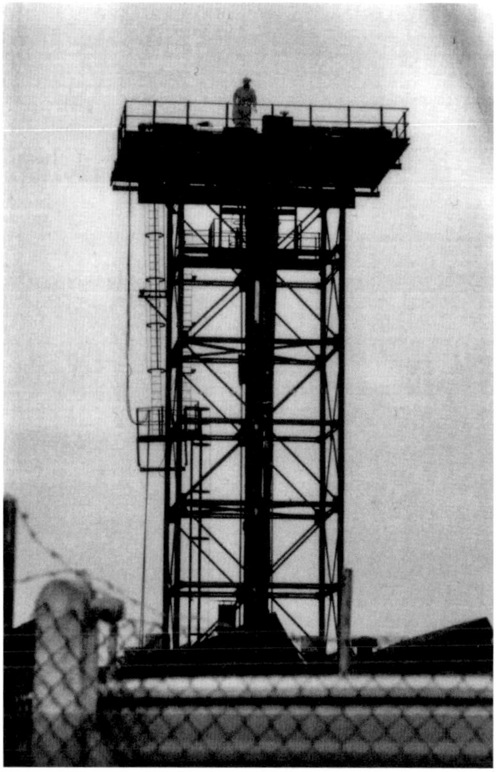

Demolition of the water tower June 2002.

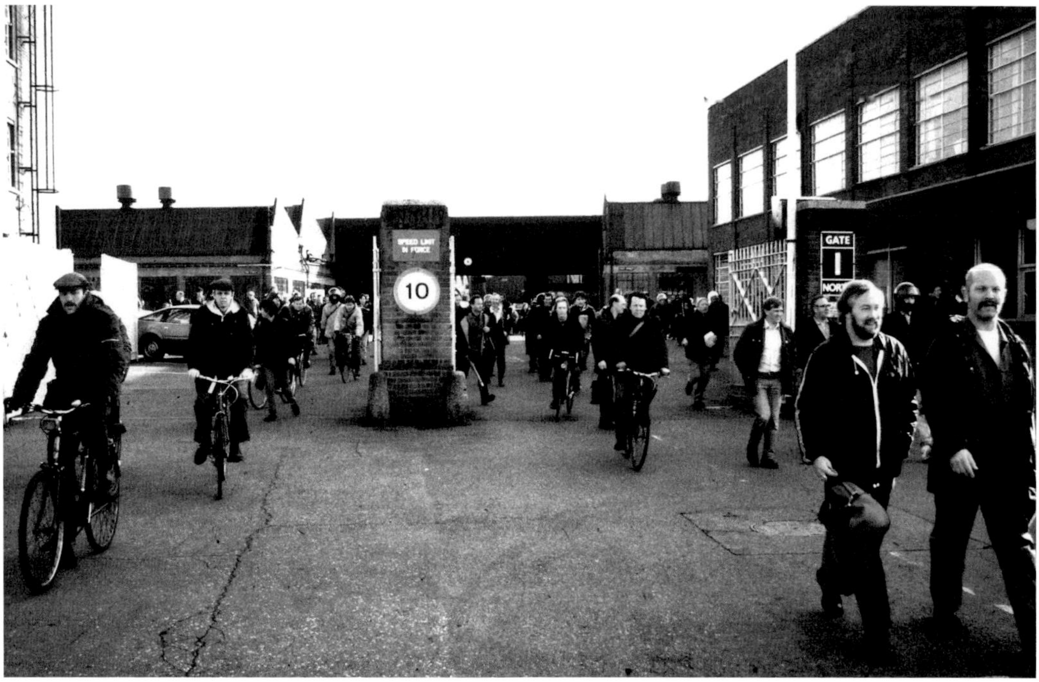

Pym's Lane gates. In the first three decades employees would be swarming through on bicycles

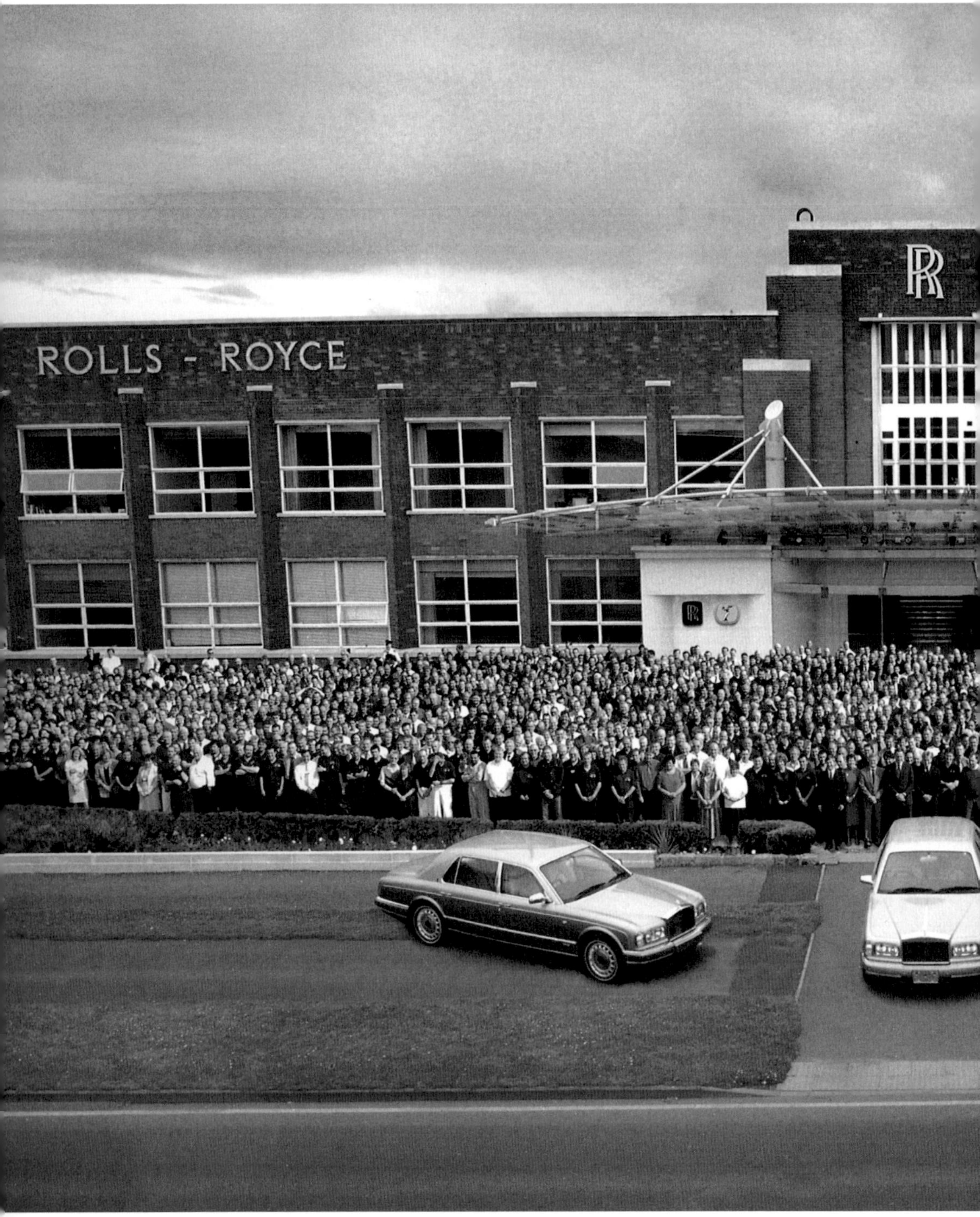

The front of the factory has been superimposed on this image as the employees were photographed on the Legends sports ground in September 2002. The three cars, a Silver Seraph, Park Ward (extended wheelbase) and Corniche were the last manufactured in Crewe by Rolls-Royce.

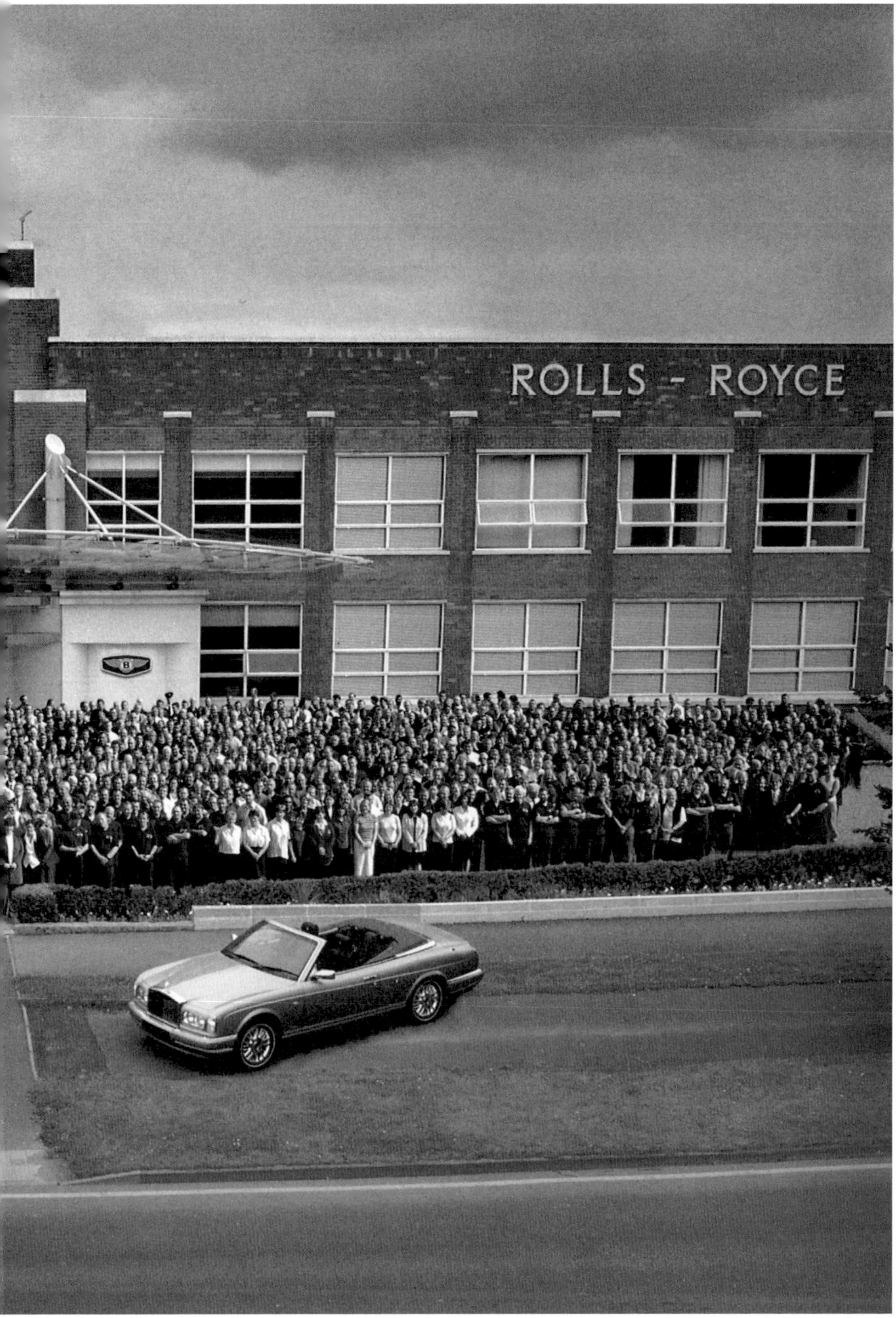
ROLLS - ROYCE

Also Available from Amberley Publishing

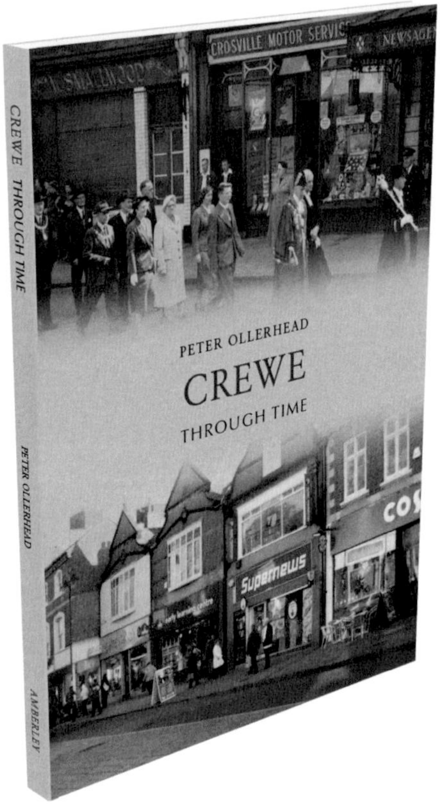

Crewe Through Time
Peter Ollerhead

ISBN: 978 1 84868 202 3
£12.99